Hope and Grief in the Anthropocene

The Anthropocene is a volatile and potentially catastrophic age demanding new ways of thinking about relations between humans and the non-human world. This book explores how responses to environmental challenges are hampered by grief for a pristine and certain past, rather than considering the scale of the necessary socioeconomic change for a 'future' world. Conceptualisations of human–nature relations must recognise both human power and its embeddedness within material relations. Hope is a risky and complex process of possibility that carries painful emotions; it is something to be practised rather than felt. As centralised governmental solutions regarding climate change appear insufficient, intellectual and practical resources can be derived from everyday understandings and practices. Empirical examples from rural and urban contexts and with diverse research participants – indigenous communities, climate scientists, weed managers, suburban householders – help us to consider capacity, vulnerability and hope in new ways.

Lesley Head is Redmond Barry Distinguished Professor and Head of the School of Geography at the University of Melbourne, Australia. This book was written while she was Director of the Australian Centre for Cultural Environmental Research, University of Wollongong, Australia.

Routledge Research in the Anthropocene

Series Editors: Jamie Lorimer and Kathryn Yusoff

The *Routledge Research in the Anthropocene Series* offers the first forum for original and innovative research on the epoch and events of the Anthropocene. Titles within the series are empirically and/or theoretically informed and explore a range of dynamic, captivating and highly relevant topics, drawing across the humanities and social sciences in an avowedly interdisciplinary perspective. This series will encourage new theoretical perspectives and highlight groundbreaking interdisciplinary research that reflects the dynamism and vibrancy of current work in this field. The series is aimed at upper-level undergraduates, researchers and research students as well as academics and policy-makers.

Hope and Grief in the Anthropocene
Re-conceptualising human–nature relations
Lesley Head

Recovering the Commons
Edited by Ash Amin and Philip Howell

Hope and Grief in the Anthropocene

Re-conceptualising human–nature relations

Lesley Head

LONDON AND NEW YORK

First published 2016 by Routledge

2 Park Square, Milton Park, Abingdon, Oxfordshire OX14 4RN
711 Third Avenue, New York, NY 10017

Routledge is an imprint of the Taylor & Francis Group, an informa business

First issued in paperback 2018

Copyright © 2016 Lesley Head

The right of Lesley Head to be identified as author of this work has been asserted by her in accordance with sections 77 and 78 of the Copyright, Designs and Patents Act 1988.

All rights reserved. No part of this book may be reprinted or reproduced or utilised in any form or by any electronic, mechanical, or other means, now known or hereafter invented, including photocopying and recording, or in any information storage or retrieval system, without permission in writing from the publishers.

Notice:
Product or corporate names may be trademarks or registered trademarks, and are used only for identification and explanation without intent to infringe.

British Library Cataloguing in Publication Data
A catalogue record for this book is available from the British Library

Library of Congress Cataloging in Publication Data
Names: Head, Lesley, author.
Title: Hope and grief in the anthropocene : re-conceptualising human-nature relations / Lesley Head.
Description: New York, NY : Routledge, 2016. | Series: Routledge research in the Anthropocene
Identifiers: LCCN 2015035844 | ISBN 9781138826441 (hardback) | ISBN 9781315739335 (e-book)
Subjects: LCSH: Human ecology. | Nature—Effect of human beings on. | Global environmental change—Social aspects.
Classification: LCC GF75.H434 2016 | DDC 304.2—dc23LC record available at http://lccn.loc.gov/2015035844

ISBN: 978-1-138-82644-1 (hbk)
ISBN: 978-1-138-54714-8 (pbk)

Typeset in Times New Roman
by Keystroke, Station Road, Codsall, Wolverhampton

Contents

List of figures vii
List of tables ix
Acknowledgements xi

1 The spectre of catastrophe 1

2 Grief will be our companion 21

3 Past, present and future temporalities 38

4 More than human, more than nature 54

5 Practising hope 74

6 Rethinking agriculture, rethinking Anthropocene 93

7 Living with weeds 115

8 Governing the ungovernable? 133

9 Beyond fortress and sprawl: retrofitting cities, suburbs and households 149

10 The Anthropoceneans 167

Index 175

Figures

1.1	Main road into Kinglake, Victoria, two weeks after Black Saturday, February 2009	1
3.1	Sand waves in the shallows of Lake Pedder with a view to the plains around Maria Lake, Tasmania, 1968	39
4.1	*FLOAT*, dancer Tess de Quincey, Siteworks 2010 at Bundanon NSW	54
6.1	*Typha* in roadside drain, Wollongong, NSW. Both old and new growth are visible	93
6.2	White plastic protects grape vines near Euston, NSW	105
6.3	Rice farms at the edge of Tuckerbil Swamp, 2012 Ramsar listed wetland, NSW	109

Tables

1.1	Contributions of cultural research to the challenges of sustainability and climate change	12
5.1	The emotional labour of distancing by climate scientists	82
6.1	Key themes in rethinking the origins of agriculture	97
9.1	Themes identified in meta-ethnography of household sustainability analyses	155

Acknowledgements

This work draws on collaborative research with a number of Australian Centre for Cultural Environmental Research (AUSCCER) colleagues: Jenny Atchison, Chris Gibson, Gordon Waitt, Nick Gill, Catherine Phillips, Natascha Klocker, Olivia Dun, Chantel Carr, Theresa Harada and Stephanie Toole. That this work has been undertaken with great professionalism and has produced richly detailed outcomes is in no small measure due to their skill, enthusiasm and generosity. None of these co-authors should be held responsible for the more speculative ends to which I have put our joint work, nor for the way I have tried to bring disparate studies into conversation with one another.

For input and discussion on the themes canvassed in the book I particularly thank Michael Adams, Harriet Bulkeley, Noel Castree, Iain Davidson, Eliza de Vet, Christine Eriksen, Carol Farbotko, Sumita Ghosh, Leah Gibbs, Ananth Gopal, Richard Hobbs, Sophie-May Kerr, Christian Kull, Brendon Larson, Helen McGregor, Heather Moorcroft, Emily O'Gorman, Matt Paterson, Jim Proctor, Priya Rangan, Joachim Regnéll, Lauren Rickards, Paul Robbins, Libby Robin, Katarina Saltzman, Gunhild Setten, Marie Stenseke, Johannes Stripple, David Trigger, Sean Ulm, Thom van Dooren, Ellen van Holstein, Justin Westgate and Mats Widgren. I am grateful to seminar audiences in Sydney, Hobart, Wollongong, Portland, Gothenburg, Stockholm and Oxford who engaged with various parts of my argument. Richard Fullagar provided domestic support and sharp insight with his usual level of generosity.

The Bundanon connection has been facilitated by Deb Ely, Mary Preece, Henry Goodall and Linden Brown.

Diane Walton and Eliza de Vet provided valuable editorial assistance.

The book draws on research funded by the Australian Research Council over a number of years (DP0211327, DP0665932, DP0986041, DP140101165). Its writing has been enabled by FL0992397.

As things have turned out, this book is being completed at the end of my time with AUSCCER. It has been the great privilege of my professional life to work with this group of people, and it has been a joy to come to work each day. I have learned a lot from all staff and students, including those not named individually here. I offer this book as a humble punctuation point in our time together, and look forward to ongoing collaborations.

Earlier drafts of some sections were published in the following papers and are reproduced with permission:

Ghosh, S. and Head, L. 2009 Retrofitting the suburban garden: morphologies and some elements of sustainability potential of two Australian residential suburbs compared. *Australian Geographer* 40: 319–346 (with permission of Taylor and Francis)

Head, L. 2011 More than human, more than nature. *Griffith Review 31: Ways of Seeing,* pp. 74–82 https://griffithreview.com/articles/more-than-human-more-than-nature/ (with permission of Griffith Review)

Head, L. 2012 Decentring 1788: beyond biotic nativeness. *Geographical Research* 50: 166–178 (with permission of Wiley-Blackwell)

Head, L. 2014 Contingencies of the Anthropocene: lessons from the 'Neolithic'. *The Anthropocene Review* 1: 113–125 (with permission of Sage)

Head, L. and Atchison, J. 2015 Governing invasive plants: policy and practice in managing the Gamba grass (*Andropogon gayanus*) – bushfire nexus in northern Australia. *Land Use Policy* 47: 225–234 (with permission of Elsevier)

1 The spectre of catastrophe

The news is not good. It feels as though we are hurtling down a hill without any brakes, through an unfamiliar landscape, to an uncertain destination. The evidence is mounting that we are well past the point where climate change response can be a planned, gradual transition. It is much more likely that profound and unwanted change in the next few years will make a mockery of current policies on climate change and other issues. We need to deal with at least the possibility of catastrophe. Yet daily life continues more or less unchanged, in varying combinations of struggle and contentment. We are in collective denial. We are grieving.

A central argument of this book is that an under-acknowledged process of grieving – with all its complexity, diversity and contradiction – is part of the

Figure 1.1 Main road into Kinglake, Victoria, two weeks after Black Saturday, February 2009. Photo: David Bruce and Bushfire and Natural Hazards CRC.

cultural politics of responding to climate change and associated environmental challenges. Specifically, I argue that grieving helps explain the denial we face and experience in accepting the scale of the changes required in ways of living. My perspective is Australian, but it is increasingly apparent that these conditions are shared across much of the affluent West (Randall 2009, Doherty & Clayton 2011). This is the converging, congealing grief at the loss of the conditions that underpin contemporary Western prosperity. It is grief for the approaching demise of the conditions sustaining life as we know it – the thing most of us did not know was called the Holocene. It is grief for the loss of a future characterised by hope. A Swedish colleague with a young child put it to me in this way: 'I would be completely OK with living a much simpler life, but it's deeper than that . . . I feel as if I can never be completely happy ever again'.

These are difficult issues to discuss in contemporary Western society. We debate climate change at length, mostly framed in the spurious terms of whether the science is settled enough for us to make some long-term decisions. But even those who know the science most intimately face strong social pressures to be optimistic about the future. There is deep cultural pressure in the West not to be 'a doom and gloom merchant'. Hence, even when the evidence points towards the strong possibility of some catastrophic scenarios, the tendency is to focus policy and action on the most optimistic end of the spectrum of possibilities. But at least some of us should be thinking systematically about worst-case scenarios. In this book I attempt to do this, engaging carefully with what the implications might be. If we have at least the possibility of catastrophic outcomes, what should our response be? I reject the cultural assumption that even to canvass these issues is to give in to them, to give up or to assume the worst. Rather I argue that a relentless cultural disposition to focus disproportionately on positive outcomes is itself a kind of denial. I argue that grief is a companion that will increasingly be with us. It is not something we can deal with and move on from, but rather something we must acknowledge and hold if we are to enact any kind of effective politics. Or, to put it differently, it needs to become an explicit part of our politics.

The challenges facing us have been well rehearsed elsewhere. Climate change is not a stand-alone issue, but interacts with long-standing negative human impacts; biodiversity loss, land and water degradation, pollution (Sutherland et al. 2014, Whitehead 2014). This combination of processes has come to define the Anthropocene, as a geological age in which human activities dominate earth surface processes. Nevertheless, the most urgent implication of climate change science is that we need to keep 60–80 per cent of the fossil fuel reserves already listed on world stock exchanges in the ground to have a chance of avoiding global warming of 2ºC (Carbon Tracker 2014). Specifically, this requires that a third of global oil, half the gas and over 80 per cent of current coal reserves need to remain unused to meet the 2ºC target (McGlade & Ekins 2015). Pricing in the risk to current investments of these 'stranded assets' would lead to a significant financial crisis. Put another way, we need to decarbonise at the rate of 9–10 per cent per year for at least a decade to avoid two degrees of warming. There is no historical analogy for how to do this; the 2008 Global Financial Crisis led to only a 1.4 per cent decrease, which was quickly reversed. If business as usual continues – and many scientists think it is already too late to avoid two degrees of warming, due to the lag effects of emissions already in the atmosphere (Anderson & Bows 2011) – then we are on track for 4–6 degrees of warming with an increase in extreme events, and fundamental changes in underlying conditions. It does not sound like much, but that is the temperature difference between now and the last ice age, in the opposite direction.

The 'unburnable carbon scenario' poses significant challenges to human survival, and thus to the scale of socioeconomic transformation that we face. The possibilities seem to include 'planned economic recession' (Anderson & Bows 2008, p. 3880) or economic collapse forced by climate change. Either way, we must imagine that drastic changes to everyday life are in the offing. Transformational rather than incremental change refers not only to the possibility of a 4°C warmer world (Stafford Smith et al. 2011, Park et al. 2012), but also to the increased level of surprise associated with rapid change in complex systems. These are terrifying thoughts, given that humans are not good at voluntary restraint, and given the way all our lives and wellbeing in the more affluent parts of the world are tied into and dependent on a fossil fuel economy. We do not yet know how much transformation will proceed deliberately and how much will be forced on us, but it is likely that we will be forced as much as governed to low carbon pathways.

This scale of transformation should shift our thinking in diverse ways, including from scarcity to abundance. For a number of decades now, we have thought of resource crises in terms of how to make non-renewable resources last longer. We were taught, and have taught our students, that the fossil fuel age must necessarily end because the resources were non-renewable. We did not know what they would be replaced with, nor when it would happen, but the exponential growth curves of the second half of the twentieth century, spinning out into the future, have always been presented as signals of impending scarcity – of food, energy, land and water, fuelled by population growth and increasing affluence. Many, if not most, of our environmental debates have been framed in terms of scarcity and running out of things – peak oil, peak phosphorous, peak lots of things, loss of biodiversity. Because of the challenge of keeping coal and oil in the ground, it now looks to be abundance, not scarcity, that we must address first, and in our own lifetimes rather than an unimaginable future time. Yet we do not really have a vocabulary, conceptual or otherwise, that links abundance and excess with environmentally good outcomes.

Of course, these issues have been around for several decades now. Re-reading overviews of climate change response such as Rayner and Malone (1998), one is struck by how much of the debate and knowledge was well established more than fifteen years ago. However, one big thing has changed, arguably due to the failure of society to deal with climate change issues when they were still potentially manageable – the sense of urgency is much stronger today, together with increasing awareness of likely non-linear changes whose specifics will be impossible to predict. In contrast, Rayner and Malone depicted the situation as still possible to deal with via deliberate and incremental change:

> In the grand scheme of things, climate change is probably not the deciding factor in whether humanity as a whole flourishes or declines. The resilience of human institutions and their ability to monitor and adapt to changing conditions seems to be more important.
>
> (1998, p. 29)

These days Rayner works on geoengineering,[1] as a potential response to worst-case scenarios. If we are at the point of systematically considering 'deliberate large-scale intervention in the Earth's natural systems to counteract climate change',[2] my contention is that we need to systematically consider the concept of catastrophe, and the way in which our socioeconomic systems could potentially unravel. In the case of geoengineering most of us would prefer to avoid such considerations, and many would be concerned by, and want to contest the hubris implicit in, the idea that humans could 'manage' the climate. In the case of a potentially catastrophic dismantling of the basis of much everyday life in the twenty-first century, we need to be able to think about this without the easy charge of 'doom and gloom merchant'. My aim here is to find a calm and hopeful way to think and talk about these painful issues.[3]

Which Anthropocene are we talking about?

There are many Anthropocenes in current discussion; Dibley (2012) has identified seven. Broadly speaking, we can distinguish between a scientific discussion over whether we have entered a new geological age, and a wider social debate in which #Anthropocene has entered the public discourse as an emblem of the way humans pervade all dimensions of the Earth. The former is currently being debated by the International Geological Commission, and the latter can be seen in social media discussions.

The discussion in this book shifts between the two arenas. I am not seeking to intervene in the geological debates over whether and when we have a golden spike, but I am interested in those debates for what they tell us about how scientists understand the human relations to the more-than-human world. The evidence of the Anthropocene requires us to rebuild its own conceptual scaffolding in order to imagine and enact the world differently (Sayre 2012). This will be a long-term project, in which the maintenance of critical perspectives is essential. It is also urgent!

Such paradoxes seem to lie at the heart of the Anthropocene concept. It challenges the ideal of economic growth that helped propel it, particularly its manifestation over the second half of the twentieth century (Steffen et al. 2011). If human impact on the earth can be translated into human responsibility for the Earth, the concept may help stimulate appropriate societal responses and/or invoke appropriate planetary stewardship (Ellis 2011, DeFries et al. 2012). Even so, while the concept has emerged out of palaeoecological, archaeological and historical perspectives on earth systems, there is great uncertainty about the future, and how we can apply any lessons of the past, since 'Earth is currently operating in a no-analogue state' (Crutzen & Steffen 2003, p. 253).

Much of this book will focus on the paradox of the human. The Anthropocene is presented as a time period defined by the activities and impacts of the human, yet it is paradoxically also a period that is now out of human control, due to rapid, unpredictable and non-linear change. Conceptualisations of human–nature relations must recognise both human power and its embeddedness within material relations. We separate out humans at the same time as the evidence shows how deeply embedded we are. Exactly what 'conceptual scaffolding' we can best use to live in a world of non-linear change is explored further below.

Several authors have argued that the emergence of the Anthropocene concept is productively a moment of convergence between 'Earth System natural science and post-Cartesian social science' (Malm & Hornborg 2014, p. 62, Lorimer 2012, Oldfield et al. 2014). This convergence is characterised by: seeing outcomes as contingent, acknowledging the demise of nature as a realm separable from culture, emphasising non-linear changes and uncertainties, and attending to the material basis of interspecies interactions including those within and between humans and others. The convergence thus provides a historical opportunity to challenge the modernist framing of humans as separate from and superior to nature, and of human history as a process of continuous improvement.

In order to make the most of this moment, it is necessary to forestall two attendant risks. The first is abandoning contingency to teleology and essentialism. The second its to reify the Anthropocene too quickly as just another phase in human history. That will not only be historically inaccurate, but also have limited potential to mobilise the kinds of political action that its constituent evidence demands. It is more likely to lead to fatalistic responses. I argue that we should use the period when the Anthropocene concept is still emergent in the public consciousness, and informal as a geological epoch, to craft an articulation that is more consistent with contingent understandings of history and science, attuned to variability and (as it happens, in the process) generative of political possibility.

So the emerging concept of the Anthropocene contains both risks and opportunities in the possibility of conceptual reframing. In this book I aim to contribute constructively to that project. In what follows I outline what I see as the key rethinking required, including about the Anthropocene itself, and explain the structure of the book.

Grief

The grieving we will discuss is twofold: first, grieving for the modern self, a process discussed in Chapter 2. Grief is manifest across society in our denial of the scale of the necessary changes to our socioeconomic underpinnings. In a sense this is grief for what we understood as our future – hitherto a time and place of unlimited positive possibility.

We are hampered in our response to these challenges by a second manifestation; grief for a stable, pristine and certain past. This is particularly the case for contemporary environmental thought, much of which has a linear framing in which the past is the baseline and everything that follows is loss. So, at a time when we face the most profound environmental challenges ever, environmental thinking that grew out of modernism finds itself under-equipped to provide new tools. In Chapter 3 I show how the pristine past continues to provide a benchmark against which many environmental ideals are measured. Loss and mourning have been an explicit part of biodiversity conservation debates *because* of their temporal reference point against a past baseline. But that reference point has not necessarily helped them deal with the grief. This chapter also identifies some of the ways environmental management is shifting. To the extent that we have any control over things, the past should only be part of how we think about environmental protection and management. We need symbols and themes that allow us to work towards possible futures as well as acknowledge a grieved-for past.

There are parallels in this thinking about stability with the way climate itself is understood, as Hulme has argued:

> And it is perhaps this ideology of wildness, the idea of world climate as the ultimate refuge of the natural, which has driven much of the thinking which lies behind the fourth example I use of how we load climate with our ideologies. If an untouched climate, a pure and natural climate, is to be valued, then maintaining its stability becomes of prime, even sacrosanct, importance. Climate thus becomes freighted with the ideology of stability and order in Nature, as opposed to ideas of change and chaos.
>
> (2009, p. 26)

Hulme does not use the word grief, but he goes on to argue that this is 'Lamenting Eden' (p. 342). In parallels with Romantic thought in Western environmentalism, 'climate becomes something that is fragile and needs to be protected or "saved"' (p. 343). The fact that the past has never been static, and the future has never been assured, is irrelevant to their nostalgic and aspirational power respectively.

Grief and mourning are generally seen as negative emotions, or at least ones we would choose not to experience if we had the choice. Often in discussion about the challenges facing the world, people are very quick to block off negative thoughts and focus on the positive. My argument here is that this switch is itself part of a collective denial and that we need to urgently find ways for society to bear, and bear witness to, the painful emotions around climate change. The argument is illustrated in Chapter 5 through a group of climate scientists. They distance themselves from stress and anxiety by downplaying the negative emotions and playing up the positive ones (love of their job, passion for science). This double move enables them to keep going, but in ways that systematically downplay worst-case scenarios and embody a kind of everyday denial – an unjustified bias towards positive scenarios. By rationalising emotions they systematically avoid the worst possibilities, even where they are as statistically likely to occur as not. This empirical study provokes important questions for all of us in a context of strong social and cultural pressure to be positive rather than negative. Rational, embodied responses suggest we will need to bear painful emotions (fear, grief, anxiety) if we are to be effective and truthful. We need to ask, in a profound cultural and psychological sense, why painful or difficult emotions might sometimes be paralysing? And can we find ways to bear such emotions without paralysis, and thus better manage the emotional labour of addressing climate change?

The Anthropocene as a modernist concept

The Anthropocene is an ostensibly radical concept, identifying humans as the dominant factor in global processes, and demanding major changes to the way we interact with the natural world. But in other ways it is more of the same, and risks perpetuating a modernist understanding of human domination over nature; 'there is a hyper-humanism which seeks to manage and ultimately master the ecological crisis' (Roelvink & Zolkos 2015, p. 47).

The emerging narrative tends to present human history in a linear, deterministic and teleological frame at odds with both scientific and social scientific understandings of evolutionary and historical contingency (De Landa 1997). The Anthropocene is presented as just another stage of human history, sometimes with a sense of inevitability about it. This may be an inadvertent by-product of the geological framing of the concept. The geological history of the Earth is one of great contingency, but the visual representations of geological history, as strata and ages, accumulating towards the present, suggest a much more progressivist metaphor. The way discussions around the Anthropocene continue a view of history as linear and teleological – as something of an inevitable stage in human history – are explored in Chapter 3. It is important to emphasise, as does De Landa (1997), that a progressivist view of the unfolding of life and human history was always inaccurate. Yet this narrative is even more at odds with the uncertainty and non-linear change – some of it potentially very rapid – that will characterise the Anthropocene.

Both the humanities and the sciences are implicated here, or should we say that both the humanities and the sciences are a product of the deep cultural distinction between humans and the rest of nature that has pervaded Western thought since classical times (Glacken 1967). Glendinning shows how the Anthropocene has been framed as an extension of classical humanism as well as modernity: 'This classical anthropology, this great humanism, which sees a natural but fundamental division between human beings and other living things, lies at the heart of the thinking about human history which tells of the special heading of man' (2000, p. 25).

Who is this we, the anthropos?

Who then is this 'we', the anthropos? Is it the human species, the undifferentiated human subject? If so, its experiences of the social and environmental changes under discussion already vary widely, to the extent that a shared human experience may be a rather small lens. And who is this 'we', who grieve for modernity? The modern subject in this discussion is all of us who live with the ideal of progress, however that is imagined. It is imagined differently in both left and right political orientations, which have worked towards different utopias, but both share the aspirations of modernity. The modern subject values autonomy and individual freedom, and connects the future with the possibility of improvement.

If we moderns have had the hope of progress and improvement, the reality has been dramatically unequal. As the histories of capitalism and colonialism have shown us, the hopes of many have been built on other people's suffering. In no country have we managed to build societies with both low per capita ecological footprints and the highest levels of human wellbeing, as measured by the human development index (Steffen et al. 2011). So we must acknowledge the Western-centrism of our catastrophic scenarios – for many people in many parts of the world, daily life is already, and has always been, infused with catastrophe and grief.

As others have argued, the 'species' concept may be a category mistake in the way we think about the Anthropocene: if 'some humans introduced steam-power *against the explicit resistance of other humans*, then it would be hard to maintain a notion of it as an expression of a species-wide "human enterprise"' (Malm 2014, p. 41, emphasis in original). For Malm and Hornborg (2014), the concept of an undifferentiated human in its impacts is impossible to reconcile with the huge historical and contemporary differentials in access to resources. Indeed, they argue:

> uneven distribution is a condition for *the very existence* of modern, fossil-fuel technology ... The affluence of high-tech modernity cannot possibly be universalized – become an asset of the species – because it is predicated on a global division of labour that is geared precisely to abysmal price and wage differences between populations.
>
> (Malm & Hornborg 2014, p. 64)

Other differentiations that similarly draw attention to more particular social and political drivers include the Capitalocene (Huber 2009, Malm 2013, Moore 2013) and the Econocene (Norgaard 2013).

To the question, 'What characterizes the Anthropocene?', Zalasiewicz et al. start their answer in the deep time of human prehistory:

> The use of tools was once thought to distinguish humans from all other animals, and among the earliest people who lived at 2Ma in Africa were *Homo habilis*, the 'handy man'. From that time, people have been modifying the Earth. For much of that human story, these changes were achieved by muscle and sinew, supplemented first by primitive tools, largely for hunting, and later by fire. Traces of humans in the Pleistocene rock record are rare, and stay rare until the Holocene.
>
> (2011, p. 836)

For the anthropos to hold at a species level, it has to encompass all of the relevant time and space of *Homo sapiens*. This it demonstrably does not do – despite widespread recognition of human influences on fire and fauna in the Pleistocene, there is not a serious suggestion that the Anthropocene is a late Pleistocene phenomenon (although note Foley et al.'s 2013 argument for a Palaeoanthropocene). Nevertheless, Zalasiewicz et al. in the quote above hark back even further, and to a genus level.

As Malm and Hornborg (2014) have shown, the long evolutionary path is a common trope in the standard Anthropocene narrative. A key component is the manipulation of fire. Even for the most common Anthropocene chronology, attached to James Watt's eighteenth-century mobilisation of the steam engine, the evolutionary precursor of fire is framed as the ultimate cause because the transition to fossil fuels in the Industrial Revolution needs to be:

> deduced from human nature. If the dynamics were of a more contingent character, the narrative of an entire species – the *anthropos* as such – ascending to biospheric supremacy would be difficult to uphold: 'the geology of mankind' must have its roots in the properties of that being.
> (Malm & Hornborg 2014, p. 63)

Consider the *cene* as well as the anthropos. In the narratives referred to above, the Anthropocene origin is located not only with a human ancestor, but also very deep in time. I agree with Malm and Hornborg that this is more by default than design, as scientists have sought a hook for a complex narrative. The linear view of history and prehistory is inadvertently embedded within the dominant modes of visual representation – timelines and stratigraphic diagrams (Head 2000). But the result is a teleological view of human history in which the (negative) outcome is inevitable, a visual trajectory further reinforced by the many exponential curves that characterise the Anthropocene (e.g. Steffen et al. 2011, Figure 1).

The evidence of humans and their processes being embedded into earth systems at all scales is widely understood to represent 'a very public challenge to the modern understanding of Nature as a pure, singular and stable domain' (Lorimer 2012, p. 593), separable and separated from humanity (Oldfield et al. 2014). Despite the claims, it seems that such a view of nature is only half dead since, as Proctor (2013, p. 90) argues, nature survives in most invocations of the Anthropocene: 'It appears typical, when confronted with the complexities that are the Anthropocene, to sharpen the conceptual boundary separating these domains [nature and culture] so as to render this complexity understandable'. Robbins and Moore (2013) go so far as to name the scientific anxiety involved as a disorder. The notion of socio-ecological systems, in which the two separate domains are now mixed, is another example of reinforcing rather than rethinking the dualism (Head 2012). It is not surprising that the human–nature dualism is so deeply embedded in the narrative, given its deep historical roots in Western thought (Glacken 1967, Sayre 2012), entanglement of the associated concept of nature in contemporary life (Castree 2015), and the fact that industrial capitalism is itself partly constitutive of both the dualisms that we now wrestle with and the Anthropocene itself (Sayre 2012, Malm & Hornborg 2014).

There are insights to be gained here from the collection of social science approaches referred to as posthumanist, an important theme of Chapter 4. These contest persistent human exceptionalism by tracing 'the materialities of interspecies interaction – including genetic, microbial, haptic, digestive and ecological connections – to demonstrate the ontological impossibility of extracting a human body, let alone intentional mind, from the messy relations of the world' (Lorimer 2012, p. 595 on Haraway 2008).

As Gibson and I (Head & Gibson 2012) have argued at greater length, there are both opportunities and challenges here. There is a major and ongoing challenge in elaborating human and non-human continuities and differences (part of which, following Lulka (2009) is to resist homogenising the non-human). As scholars we need to be eternally vigilant in applying the analytical impulse to questions of human difference and power, and the ways they are conceptualised in climate change debates. Plumwood's (1993) analysis of the deep structures of mastery buried in our intellectual frameworks is still apposite, and her theory of mutuality, which acknowledges both continuity and (non-hierarchical) difference between humans and non-humans, continues to be helpful here. And of course it is in some ways an inescapable dilemma; 'Our life condition appears to be "both/and" rather than "either/or", obliging us to use the contradictory ideas of nature as "external" and "universal" when discussing ourselves' (Castree 2015, p. 29). A key point is that these debates and tensions are a fundamental aspect of how and whether we conceptualise the Anthropocene, not concerns to be sidelined as a simple definitional footnote.

So, the anthropos at the core is a slippery concept – at once too large and too separated to really understand what is going on. The divide between anthropos and other (usually thought of as humans and nature) is one of the many connected dualisms that must be undone and rethought.

Let me be clear – this is not an argument to get rid of the concept of the human, but to consider more carefully differentiations of concept and practice both within this category, and between it and others. We have to think differently about how human and other life and materials are mutually embedded, while at the same time accounting for clear evidence of different power relations within such assemblages (Head & Gibson 2012). The challenge of rethinking the human in associationist rather than separationist terms (see also Glendinning 2000, Anderson 2007) is addressed in Chapter 4. In the process we can start to re-imagine humans as a force for environmental good, or at least not essentially bad or damaging, as some of the Anthropocene framing suggests.

This should not involve undoing dualisms to reinsert others – the most notable of recent binaries is that between the good and bad Anthropocene. I disagree with Lynas (2011, p. 22) who goes from the naturalness of humans to the inevitability of technical solutions. It is my belief that the evidence is actually much messier than that, and so the categorisations should be messed up. Like Leduc

(2014, p. 248) I want to ask, 'What if a sustainable response to climate change requires questioning modernity's still persistent implicit ontology by rendering "explicit the prevailing conception of being" and how it is maladaptive in the context of an interconnected reality'?

Who is this we, the reader?

The book is situated firmly as a social science contribution to debates around the Anthropocene and climate change. In pointing to many points of intersection with the natural sciences, I hope it will also find a readership there. The 'we', the reader and writer together, are usually from the affluent parts of the world who have benefited so clearly from the fossil-fuel economies that now constitute our most pressing problem. To be able to have these conversations presumes much; in particular we have presumed an expectation of happiness, or at least its possibility. That is not a universal expectation, and for many people in the world the present day is already and always a situation that we would call catastrophic. Being beneficiaries in such a drastically unequal world brings responsibilities to help construct things otherwise.

Hope

The 'hope' in the title is messy, fraught and uncertain. The argument I make about hope, and where it might be found, consists in decoupling it from the emotion of optimism. Hope savours the life and world we have, not the world as we wish it to be. If the relationship between grief and denial challenges us to acknowledge and bear negative emotions without becoming paralysed, the corollary is that we should not depend on positive emotions (e.g. optimism) to provide the basis for hope. The concept of hope I advance here, drawing on various attempts to theorise hope in human geography, is thus found in practices rather than particular emotions.

Hope can also be found in unexpected places. An important source of this gritty, keeping-going kind of hope is found in diverse sets of vernacular practices, as outlined in Chapters 5 to 9. The broad argument here is that widening our horizons to diverse practices allows us to imagine alternative possibilities, in the way outlined by Gibson-Graham (2008), and to identify the existing cultural resources that we have to facilitate change. De Landa thinks about this in a related way, by suggesting that profound historical change tends to come from the points of flexibility outside the concentrations of power: 'the creation of *novel* hierarchical structures through restratification is performed by the most destratified element of the *previous* phase' (1997, p. 266). Further, in a context of distributed agency and non-linear change, it is impossible to predict where all the good or bad things will happen, or even which processes and states are likely to be more adaptive (or not). As Hulme (2009, p. 84) argues, for most of human history we have assumed uncertainty, whereas now we often think of it as an obstacle to collective action.

In a context where we have to keep the coal in the ground, unseat fossil capitalism and remake our modernist identities, Chapters 5 to 9 talk about some apparently very small things. What is the purpose of dealing with these apparently 'small' issues of culture (sometimes individual, sometimes household, sometimes society-wide) when revolution is needed? The contribution of in-depth qualitative and ethnographic research on which these chapters draw often needs more explanation and defence than that of quantitative methods (Head et al. 2005), and the findings do not necessarily sit easily with policy connections (Adger et al. 2013). An important strength is that such methods allow in-depth approaches to the everyday, and to common sense or taken for granted understandings and practices. The potential contributions of this kind of research are summarised in Table 1.1.

Table 1.1 Contributions of cultural research to the challenges of sustainability and climate change

1 Showing how concepts and categories matter
- Conceptual critique of problem framing
- Showing how categories and concepts (e.g. nature, nativeness, climate change) become embedded in societal structures, legislation, institutions and public discourse

2 Reframing human–environment relations
- Analyse how the 'human' and the 'social' are conceptualised in different framings (e.g. Anthropocene, ecosystems, earth systems, socio-ecological systems)
- Analyse in what circumstances humans should be understood as being apart from the rest of the natural world, and/or embedded as part of it
- Indigenous and other non-Western ontologies offer a range of cultural resources in the necessary reframing of relationships between humans and their environment

3 Providing social context
- Identify that the matter of priorities and objectives is a cultural and political issue rather than a scientific one
- Understand cultural variability and different world views in how climate is conceptualised and experienced
- Analyse climate change scepticism and denial as significant phenomena requiring cultural analysis
- Identify and clarify conflict around environmental issues
- Better understand the distributive dimensions of global environmental problems, including the unequal distribution of resources and power

4 Using depth to understand contradiction and paradox
- Complement to quantitative methods
- Explanations of paradox and contradiction identified with quantitative methods

5 Documenting variability
- Understanding the environmental cultures of different sectors of the community, and how these intersect
- Historical, class, gender, ethnic and spatial dimensions of different issues
- Diverse environmental knowledges
- Identification of cultural capacity deficits in existing natural resource management institutions and practices

6 **Understanding environmental norms and practices**
   - Critical and historical analysis of what stands for common sense and normal behaviours (e.g. cleanliness, waste) and how these came to be
   - Role of everyday routines, habits, emotions and practices in shaping environmental impacts and adaptive capacities
7 **Identifying cultural resources and thresholds for change**
   - Resources for imagining alternative ways of doing things
   - Resources for change can emerge from unlikely places (e.g. older people who explicitly do not identify as 'green' but who nevertheless have a number of sustainable practices based on frugality)
   - Alternatives can emerge from more diverse scales of governance than the national and international (e.g. household)

Source: Created using data from Head and Stenseke (2014).

At a time when top-down intergovernmental action seems not to be up to the task, survival may depend on more localised vernacular understandings and practices. Important intellectual resources come from places understood as marginal to environmental preservation; indigenous engagements, gardens, suburbs, farms, domestic homes. We can revisit empirical evidence from these to consider capacity and vulnerability in new ways. If everything spirals into catastrophe, practices imagined as sustainability practices can be reframed as survival skills (Gibson et al. 2015).

These examples help us discuss abundance and excess as well as scarcity (Chapters 6 and 9). They contribute to ontological reframing, of how to think things differently; Chapter 6 uses the concept of agriculture. I show how people change their understandings – through practice – of very binary concepts such as nativeness, and learn to live with weeds (Chapters 7 and 8). In Chapter 9 I argue from our collaborative work on households that green subjectivities are not the point – we do not need to direct people into what we understand as 'environmental' thinking. Rather we can identify and celebrate everyday practices that are associated with other subjectivities, but contribute through, for example, frugality and a hatred of wasting things.

It should be clear from all this that I do not understand climate change and the dilemmas of the Anthropocene as single processes that demand single, mega solutions, whether for example via nuclear power or rapid reduction in global population (through as yet unspecified means). It is important not to gloss over the complexities in operationalising these or any other 'magic bullets'. As a number of authors have pointed out, there is a tendency to elide the politics implicit in their use (Hornborg 2014, Swyngedouw 2014, Taylor 2015).

The Australian contribution

The examples in the second half of the book are taken from different collaborative projects I have worked on over the last few years; climate scientists (Chapter 5), debates over Australian prehistoric and contemporary agriculture (Chapter 6), the social lives of invasive plants (Chapters 7 and 8) and sustainability and climate change in households and suburbs (Chapter 9).[4] On the face of it this is an eclectic mix of subject matter, and risks the reflection of self discussed by Sara Ahmed:

> To name one's archive is a perilous matter; it can suggest that these texts 'belong' together, and that the belonging is a mark of one's own presence. What I offer is a model of the archive not as the conversion of self into a textual gathering, but as a 'contact zone'.
>
> (2004, p. 14)

Three things hold this particular 'contact zone' together in a state of creative friction. First, it was assembled over time through projects that sought to understand the cultural dimensions of human/environment relationships in diverse contexts. This central geographic concern with the way in which humans interact – both materially and conceptually – with the more-than-human world connects both prehistoric and contemporary timescales. Second, some of the studies have an empirical focus on plants and their human entanglements. This has been a particular interest shared with colleague Jenny Atchison. It provides an insight into different human lifeways, and has also led us into exploring questions of human–plant difference and otherness.

The details of course could have been otherwise. Colleagues are examining equally relevant issues in Australia with similar methods; for example the future of making and manufacturing (Gibson & Warren 2014, Carr & Gibson 2015), human relations with oceans (Gibbs & Warren 2015) and animals (Adams 2013), and preparedness and management of bushfire (Eriksen 2014, Gill et al. 2015).

So the third and perhaps the strongest thing that holds this contact zone together is its situatedness in and from Australia. Australia is an affluent Minority World country but with a distinctive colonial heritage that still infuses contemporary society, notwithstanding the huge ethnic diversity that has accompanied its migrant experience. It is usefully a combination of the Centre and the Other. As Gibson and I (in press) have argued, Australia is 'an exemplary place to contemplate everyday life on an increasingly volatile earth'. The Australian continent is arid and fire-prone, with low soil fertility and low relief. It has a diversity of habitats – tropical, temperate, arid and alpine – and its climate regimes are characterised by high variability. Australia occupies a fraught place in contemporary climate change debates and practices; historical dependence on fossil fuels such as coal and iron ore has given us high per capita greenhouse gas emissions. We are both strong contributors to climate change and particularly vulnerable to its impacts. At the same time, a combination of factors is keeping us in the kind of collective denial discussed in this book. These include vested financial interests, recalcitrant governments and the national myth of vernacular stoicism – the capacity to cope with 'droughts and flooding rains'[5] – in the face of environmental adversity.

We still have work to do in coming to terms with the rage and grief of Australia's colonial heritage. But if we misdirect this work as nostalgia for a lost paradise we are also disabling constructive engagement with our future. If Australia has particular problems with binaries we may also have the means to move beyond them, arguably to make a contribution to reframing modernity that will extend beyond our shores. In our everyday engagements with the messiness of our cultural and ecological hybridity we are all sowing the seeds of that crop. We do not yet know what it will look like, but its unruly possibilities might enhance the practice of hope.

The Australian experience and engagement have in the past been constitutive of particular views of international history; Anderson (2007) shows the role of the colonising encounter with the Indigenous Australian in challenging nineteenth-century understandings of the human. She ponders further whether Australia can thus be constitutive of a radical undoing. It is something of a stretch to argue that Australian evidence from the twenty-first century might similarly challenge persistently binary framings of the human, and contribute to a more positive framing of the human for the challenges of the Anthropocene. More modestly and accurately, all perspectives come from somewhere (to paraphrase Donna Haraway (2008)), and we need to consider this context, particularly against grand and universalising claims such as those of the good or bad Anthropocene. We can make positive contributions to destabilising unduly normative concepts by illustrating the depth of their colonial heritage and embedding – the concept of nativeness in relation to plants and animals is one example here. We can illustrate cultural diversity as a resource for imagining things otherwise.

The specificity of my examples reminds us that the Anthropocene is always somewhere, that it emerged at different times and that it will take different expression in different places. It is important to be alert to spatial and temporal variability, and what it means for phases of history. The process by and rates at which both agriculture and the Anthropocene became global in scale are clearly matters for ongoing empirical analysis. The point is that detailed analysis of such change is important for understanding causal processes, for example in disentangling drivers and effects, and imagining how and where to intervene. For example Steffen et al. (2011) clarify the possibilities of intervention with their demonstration that the post-1950 Great Acceleration 'was disproportionately driven by consumption patterns in the Global North, *even* in the context of increased population growth throughout the rest of the world' (Ogden et al. 2013, p. 342, emphasis in original). This invites interventions around consumption rather than population per se. It is particular groups of humans doing particular things that generate particular historical processes, in assemblage or constellation (Ogden et al. 2013) with many non-human others, whether we are talking about Pleistocene fire and megafaunal hunting, methane emissions from rice agriculture in China, Watt's steam engine and the parallel engines of industrialisation and colonisation, or the post World War II acceleration.

Generating political possibility

We are living in the Anthropocene as we work on it. We necessarily have to work all this out as we go along, only partially with hindsight. We are discussing a category, built out of a body of evidence, which demands that we also engineer political, social and economic change. As it happens, a contingent, messy, non-linear view will likely serve us better politically, given the failure so far of large governance categories such as nation states and intergovernmental agreements to curb emissions.

In the concluding chapter I tie the threads of the previous arguments together, and summarise some of the likely characteristics of citizens of the Anthropocene, the Anthropoceneans. As scholars we are in and of this history, and need to attend to the processes of category and thought construction just as much as the historical evidence of concern. A more contingent understanding of the Anthropocene is not only more historically accurate, it also provides more realistic and less fatalistic pathways to the future. If we are assuming humans will be part of the future, how can we articulate and enact the necessary creative human interventions – the creative destruction of dismantling the fossil-fuel economy, and a variety of restoration and repair activities? It may be out of the practice of these interventions that new concepts and practices of the anthropos emerge.

Notes

1 The Oxford Geoengineering Programme (OGP) (www.geoengineering.ox.ac.uk/) has developed a set of principles ('The Oxford Principles' www.geoengineering.ox.ac.uk/oxford-principles/principles/) intended to guide the development of geoengineering techniques. These include, for example, that 'any decisions with respect to deployment should only be taken with robust governance structures already in place, using existing rules and institutions wherever possible'.
2 This is the definition of 'geoengineering' used by the OGP (see website in above note).
3 My take on these issues has some things in common with cultural responses such as the Dark Mountain Project, 'a network of writers, artists and thinkers who have stopped believing the stories our civilisation tells itself. We see that the world is entering an age of ecological collapse, material contraction and social and political unravelling, and we want our cultural responses to reflect this reality rather than denying it' (http://dark-mountain.net/about/the-dark-mountain-project/). However my discussion is more with those who are choosing to remain engaged with the mainstream rather than withdrawing from it.
4 Most of these projects have been collaborative. The publications from these projects are cited throughout the text, where the co-authors can also be identified. In this book I extract some key examples from different projects and attempt to bring them into conversation with one another. To enhance the narrative flow, details of methods are removed to endnotes. Further information about the original data and interpretive discussion can be found in the referenced publications.
5 Dorothea Mackellar was an Australian poet best remembered for her poem 'My Country'. It contains the following lines:

> I love a sunburnt country,
> A land of sweeping plains,
> Of ragged mountain ranges,
> Of droughts and flooding rains.
> I love her far horizons,
> I love her jewel sea.

>   Her beauty and her terror –
>   The wide brown land for me!
>   (www.dorotheamac
>   kellar.com.au/)

The description of Australia as a land 'of droughts and flooding rains' is often used in public discourse to normalise climatic variability. It is arguably in increasing use to deny the significance of anthropogenic climate change. See further discussion in Chapter 6.

References

Adams, M. 2013. "'Redneck, barbaric, cashed up bogan? I don't think so': Hunting and nature in Australia." *Environmental Humanities* 2: 43–46.
Adger, W. N., J. Barnett, K. Brown, N. Marshall and K. O'Brien. 2013. "Cultural dimensions of climate change impacts and adaptation." *Nature Climate Change* 3: 112–117.
Ahmed, S. 2004. *The Cultural Politics of Emotion*. New York: Routledge.
Anderson, K. 2007. *Race and the Crisis of Humanism*. London: UCL Press.
Anderson, K. and A. Bows. 2008. "Reframing the climate change challenge in light of post-2000 emission trends." *Philosophical Transactions of the Royal Society A* 366: 3863–3882.
Anderson, K. and A. Bows. 2011. "Beyond 'dangerous' climate change: Emission scenarios for a new world." *Philosophical Transactions of the Royal Society A* 369: 20–44.
Carbon Tracker. 2014. "Unburnable carbon 2013: Wasted capital and stranded assets." Accessed 26 October 2014. Available at www.carbontracker.org/report/wasted-capital-and-stranded-assets/.
Carr, C. and C. Gibson. 2015. "Geographies of making: Rethinking materials and skills for volatile futures." *Progress in Human Geography*, DOI: 10.1177/0309132515578775.
Castree, N. 2015. *Making Sense of Nature: Representation, Politics and Democracy*. Abingdon: Routledge.
Crutzen, P. J. and W. Steffen. 2003. "How long have we been in the Anthropocene era?" *Climate Change* 61: 251–257.
De Fries, R. S., E. C. Ellis, F. S. Chapin III, P. A. Matson, B. L. Turner II, A. Agrawal, P. J. Crutzen, C. Field, P. Gleick, P. M. Kareiva, E. Lambin, D. Liverman, E. Ostrom, P. A. Sanchez and J. Syvitski. 2012. "Planetary opportunities: A social contract for global change science to contribute to a sustainable future." *BioScience* 62(6): 603–606.
De Landa, M. 1997. *A Thousand Years of Nonlinear History*. New York: Zone Books.
Dibley, B. 2012. "'The shape of things to come': Seven theses on the Anthropocene and attachment." *Australian Humanities Review* 52: 139–153.
Doherty, T. J. and S. Clayton. 2011. "The psychological impacts of global climate change." *American Psychologist* 66(4): 265–276.
Ellis, E. C. 2011. "Anthropogenic transformation of the terrestrial biosphere." *Philosophical Transactions of the Royal Society A* 369: 1010–1035.
Eriksen, C. 2014. *Gender and Wildfire: Landscapes of Uncertainty*. New York: Routledge.
Foley, S. F., D. Gronenborn, M. O. Andreae, J. W. Kadereit, J. Esper, D. Scholz, U. Pöschl, D. E. Jacob, B. R, Schöne, R. Schreg, A. Vött, D. Jordan, J. Lelieveld, C. G. Weller, K. W. Alt, S. Gaudzinski-Windheuser, K.-C. Bruhn, H. Tost, F. Sirocko and P. J. Crutzen. 2013. "The Palaeoanthropocene – The beginnings of anthropogenic environmental change." *Anthropocene* 3: 83–88.

Gibbs, L. and A. Warren. 2015. "Transforming shark hazard policy: Learning from ocean-users and shark encounter in Western Australia." *Marine Policy* 58: 116–124.

Gibson, C. and L. Head, In press. "Anticipating forced change and extremity: A postcard from the Australian pivot of geocapitalism." In *Capitalism and the Earth*, edited by N. Clark, A. Saldanha and K. Yusogff. New York: Punctum Books.

Gibson, C., L. Head and C. Carr. 2015. "From incremental change to radical disjuncture: Rethinking everyday household sustainability practices as survival skills." *Annals of the Association of American Geographers* 105(2): 416–424.

Gibson, C. and A. Warren. 2014. "Making surfboards: Emergence of a trans-Pacific cultural industry." *Journal of Pacific History* 49(1): 1–25.

Gibson-Graham, J. K. 2008. "Diverse economies: Performative practices for 'other worlds'." *Progress in Human Geography* 32(5): 613–632.

Gill, N., O, Dun, C. Brennan-Horley and C. Eriksen. 2015. "Landscape preferences, amenity, and bushfire risk in New South Wales, Australia." *Environmental Management* 56: 738–753.

Glacken, C. J. 1967. *Traces on the Rhodian Shore*. Berkeley: University of California Press.

Glendinning, S. 2000. "From animal life to city life." *Angelaki: Journal of the Theoretical Humanities* 5(3): 19–30.

Hamilton, C. 2010. *Requiem for a Species: Why We Resist the Truth About Climate Change*. Sydney: Allen and Unwin.

Haraway, D. J. 2008. *When Species Meet*. Minneapolis: University of Minnesota Press.

Head, L. 2000. *Second Nature: The History and Implications of Australia as Aboriginal Landscape*. New York: Syracuse University Press.

Head, L. 2012. "Conceptualising the human in cultural landscapes and resilience thinking." In *Resilience and the Cultural Landscape: Understanding and Managing Change in Human-Shaped Environments,* edited by T. Plieninger and C. Bieling, 65–79. Cambridge: Cambridge University Press.

Head, L. and C. Gibson. 2012. "Becoming differently modern: Geographic contributions to a generative climate politics." *Progress in Human Geography* 36(6): 699–714.

Head, L. and M. Stenseke.2014. "Seven contributions of cultural research to the challenges of sustainability and climate change." *Conversations with AUSCCER.* Accessed 13 November 2014. Available at www.uowblogs.com/ausccer/2014/11/13/seven-contributions-of-cultural-research-to-the-challenges-of-sustainability-and-climate-change/#more-2533. Published 13 November 2014. (Published in Swedish as Head, L. and Stenseke, M. 2014 Humanvetenskapen står för djup och förståelse In *Hela vetenskapen! 15 forskare om integrerad forskning,* edited by E. Mineur and B. Myrman, 26–33. Stockholm: Vetenskapsrådet.

Head, L., D. Trigger and J. Mulcock. 2005. "Culture as concept and influence in environmental research and management." *Conservation and Society* 3(2): 251–264.

Hornborg, A. 2014. "Why solar panels don't grow on trees: Technological utopianism and the uneasy relations between Marxism and ecological economics." In *Green Utopianism: Perspectives, Politics and Micro-Practices*, edited by K. Bradley and J. Hedrén, 76–97. New York: Routledge.

Huber, M. T. 2009. "Energizing historical materialism: Fossil fuels, space and the capitalist mode of production." *Geoforum* 40(1): 105–115.

Hulme, M. 2009. *Why We Disagree About Climate Change: Understanding Controversy, Inaction and Opportunity*. Cambridge: Cambridge University Press.

Leduc, T. B. 2014. "Climates of ontological change: Past wisdom in current binds?" *WIREs Climate Change* 5(2): 247–260.

Lorimer, J. 2012. "Multinatural geographies for the Anthropocene." *Progress in Human Geography* 36(5): 593–612.

Lulka, D. 2009. "The residual humanism of hybridity: Retaining a sense of the earth." *Transactions of the Institute of British Geographers* 34(3): 378–393.

Lynas, M. 2011. *The God Species: Saving the Planet in the Age of Humans*. London: Fourth Estate.

Malm, A. 2013. "Steaming into the Capitalocene." At the *Institute of British Geographers Conference*, 28–30 August. Royal Geographical Society, London.

Malm, A. 2014. *Fossil Capital: The Rise of Steam-Power in the British Cotton Industry, c. 1825–1848, and the Roots of Global Warming*, PhD Thesis. Lund: Lund University.

Malm, A. and A. Hornborg. 2014. "The geology of mankind? A critique of the Anthropocene narrative." *The Anthropocene Review* 1: 62–69.

McGlade, C. and P. Ekins 2015. "The geographical distribution of fossil fuels unused when limiting global warming to 2°C." *Nature* 517: 187–190.

Moore, J. W. 2013. "Anthropocene, Capitalocene and the myth of industrialization II." *World-Ecological Imaginations: Power and Production in the Web of Life*. Accessed 17 January 2014. Available at https://jasonwmoore.wordpress.com/2013/06/.

Norgaard, R. B. 2013. "The Econocene and the delta." *San Francisco Estuary and Watershed Science* 11(3): 1–5.

Ogden, L., N. Heynen, U. Oslender, P. West, K.-A. Kassam and P. Robbins. 2013. "Global assemblages, resilience, and Earth stewardship in the Anthropocene." *Frontiers in Ecology and the Environment* 11(7): 341–347.

Oldfield, F., A. D. Barnosky, J. Dearing, M. Fischer-Kowalski, J. McNeill, W. Steffen and J. Zalasiewicz. 2014. "The *Anthropocene Review*: Its significance, implications and the rationale for a new transdisciplinary journal." *The Anthropocene Review* 1(1): 3–7.

Park, S. E., N. A. Marshall, E. Jakku, A. M. Dowd, S. M. Howden, E. Mendham and A. Fleming. 2012. "Informing adaptation responses to climate change through theories of transformation." *Global Environmental Change* 22(1): 115–126.

Plumwood, V. 1993. *Feminism and the Mastery of Nature*. London: Routledge.

Proctor, J. 2013. "Saving nature in the Anthropocene." *Journal of Environmental Studies and Sciences* 3(1): 83–92.

Randall, R. 2009. "Loss and climate change: The cost of parallel narratives." *Ecopsychology* 1(3): 118–129.

Rayner, S. and E. L. Malone. 1998. *Human Choice and Climate Change: Vol. 4 What Have We Learned?*. Columbus, OH: Battelle Press.

Robbins, P. and S. A. Moore. 2013. "Ecological anxiety disorder: Diagnosing the politics of the Anthropocene." *Cultural Geographies* 20: 3–19.

Roelvink, G. and M. Zolkos 2015. "Affective ontologies: Post-humanist perspectives on the self, feeling and intersubjectivity." *Emotion, Space and Society* 14: 47–49.

Sayre, N. F. 2012. "The politics of the Anthropogenic." *Annual Review of Anthropology* 41: 57–70.

Stafford Smith, M., L. Horrocks, A. Harvey and C. Hamilton. 2011. "Rethinking adaptation for a 4°C world." *Philosophical Transactions of the Royal Society A* 369: 196–216.

Steffen, W., J. Grinevald, P. Crutzen and J. McNeill. 2011. "The Anthropocene: Conceptual and historical perspectives." *Philosophical Transactions of the Royal Society A* 369: 842–867.

Sutherland, W. J., R. Aveling, T. M. Brooks, M. Clout, L. V. Dicks, L. Fellman, E. Fleishman, D. W. Gibbons, B. Keim, F. Lickorish, K. A. Monk, D. Mortimer, L. S. Peck, J. Pretty, J. Rockstrom, J. P. Rodriguez, R. K. Smith, M. D. Spalding, F. H. Tonneijck and A. R. Watkinson. 2014. "A horizon scan of global conservation issues for 2014." *Trends in Ecology and Evolution* 29: 15–22.

Swyngedouw, E. 2014. "Anthropocenic politicization: From the politics of the environment to politicizing environments." In *Green Utopianism: Perspectives, Politics and Micro-Practices*, edited by K. Bradley, and J. Hedrén, 23–37. New York: Routledge.

Taylor, M. 2015. *The Political Ecology of Climate Change Adaptation: Livelihoods, Agrarian Change and the Conflicts of Development*. Abingdon: Routledge.

Whitehead, M. 2014. *Environmental Transformations: Geography of the Anthropocene*. Abingdon: Routledge.

Zalasiewicz, J., M. Williams, A. Haywood and M. Ellis.2011. "The Anthropocene: A new epoch of geological time?" *Philosophical Transactions of the Royal Society A* 369: 835–841.

2   Grief will be our companion

In October 2013, when a frighteningly early start to the bushfire season saw nearly 200 homes destroyed in the Blue Mountains, west of Sydney, Australian Greens politician Adam Bandt created controversy when he drew attention to the connections with climate change. He took to Twitter and the newspapers to decry the recently elected government's proposal to dismantle Australia's new carbon pricing scheme, drawing attention to the (scientifically well-established) link between climate change and higher bushfire risk.

Not while people are grieving, said many people, from the prime minister down. 'Indecent. Insensitive', thundered the radio shock jocks.[1] Even many who agreed with Bandt wondered at his timing, while hundreds of people were still fighting the blazes. Many residents were waiting anxiously in evacuation centres, trying to grab some hours of uncomfortable sleep, not yet permitted to return to see if their homes were still standing. My colleague Christine Eriksen, who was part of a team that went in to examine the damage, cautioned against furling any wider debates into anything, just at that moment.[2]

The debate around the Blue Mountains fires sought to separate grief and climate change, as if they were both procedurally and ontologically distinct and could be treated in linear fashion. In this chapter I argue that, on the contrary, grief and climate change are inextricably entwined. Grief is one of a number of negative emotions swirling around the changes we face. As Cunsolo Willox has argued, the silence on grief and mourning 'is a serious gap in academic literature, political practice, and media discourse around climate change, and it does not match the lived experiences of people around the globe' (2012, p. 141). We need to learn how to have grief as a companion. In this chapter I build an argument that we in the affluent West are grieving for the loss of the modern self, and its dependence on a future characterised by hope. While there are some quite specific manifestations of grief that we need to acknowledge and can learn from, they swirl into something bigger. Rose encapsulated this as Anthropocene Noir – 'the story without a known ending; the looming sense of fatality; the creeping awareness that nothing can be put right' (2013, p. 215). There is a lurking anxiety around. While the central argument of the chapter is that these emotions need to be expressed and discussed more openly – that our distress needs to be held, and borne – we do return at the end of the chapter to consider the productive potential of 'the work of mourning' (Cunsolo Willox 2012).

On a personal level I am quite unqualified to talk about grief. I have yet to lose a parent, spouse, sibling or child. I am only too aware that many people in my immediate and more distant communities struggle with grief and trauma every day, not a distant grief for lives not yet even emergent. Aboriginal friends and colleagues in northern Australia attend a funeral every fortnight or so, often as the result of suicide. I am however grieving for modernity, for a future that was always foggy but was presumed to contain the seeds of positive possibility. I may not live to know the interim outcomes of the human experiment, but my children probably will, and if I have grandchildren, they definitely will. In making this argument, I am very aware that it is common for commentators, or conversationalists, to be accused of being too pessimistic. I acknowledge the risk of conveying 'pessimism's bitter pleasure',[3] notwithstanding that in everyday life I am a glass half full person. I am as split in my emotions and my head space as everyone else – I cannot quite believe that it is going to be as bad as the evidence suggests. And yet, it is impossible to get away from that evidence. We all have that mental split, even as we contemplate our own deaths in the existential crisis of everyday life.

Emotions and reason

Our reluctance to discuss these matters is complex. Calls for attention to the emotional dimensions of climate change – and their relationship to the scientific debates – are relatively recent (Farbotko & McGregor 2010, Cunsolo Willox 2012). Negative emotions create discomfort, and it is understandable that they are avoided. However, Farbotko and McGregor (2010) noted the destabilising moment of possibility that was created when a (white, male) member of the Tuvalu delegation wept in the plenary of the Copenhagen COP 15 climate change negotiations. The tears cut across this highly scripted, 'rational', political encounter. Farbotko and McGregor argued that 'the ways in which science and emotion combine with reformist or radical potential, need our attention' (Farbotko & McGregor 2010, p. 164). Their point resonates with Ahmed's argument that 'we must persist in explaining why our anger is reasonable, even in the face of others who use this anger as evidence of poor reason' (Ahmed 2004, p. 177).

Feminists have long argued that the emotional is missing in many accounts of the world (Bondi 2005). Feminist analyses of emotion have shown how it is viewed as separate from and '"beneath" the faculties of thought and reason. To be emotional is to have one's judgement affected: it is to be reactive rather than active, dependent rather than autonomous' (Ahmed 2004, p. 3). In the history of evolutionary thinking, emotions have been understood as primitive and corresponding to the animal-like stage of evolution (Ahmed 2004, p. 3). Thus a physiological view of emotion has been as the antithesis of rationality. More recent work challenges such a strict dichotomy. Roeser (2012) cites an extensive philosophical and psychological literature to argue 'that we need emotions in order to be practically rational' (p. 1035), for example in how people might choose a strategy of denial to deal with fear and anxiety. In the context of women's personal safety, Koskela argued that 'reasoning can be thought of as a strategy to contain fear, but also as a way of maintaining sufficient courage to avoid fear and, thus, to live without it' (Koskela 1997, p. 306). What is significant here is how people deal with emotion in rational ways, making practical choices about the best way to deal with experiences of fear, anxiety, love and pleasure. Conversely, strong parallels between masculinism, rationality and optimism have been identified by Rickards et al. (2014) in their review analysis of the responses of senior decision-makers to climate change. Yet, as McKinnon (2014, p. 31) says, 'It is not obvious that people in despair about tackling climate change are making a mistake.'

There have been diverse views about the effectiveness of emotions in engendering climate change action. Roeser (2012) is among those who argue that we should take emotions more seriously in the climate change debate, because 'emotional engagement leads to a higher degree of motivation than a detached rational stance on climate change' (p.1033). Painful or difficult emotions are generally, but not always, seen as paralysing (O'Neill and Nicholson-Cole 2009, Pahl et al. 2014, Cook and Balayannis 2015). They need to be assessed in a cultural context where 'the shared norm of positive talk' (Rickards et al. 2014, p. 760) is strong. I have come to the view that our reluctance to engage with or even discuss these negative emotions is becoming problematic, and helping entrench our collective denial.

Notwithstanding the broader cultural pressure to focus on perceived positives rather than negatives, the diverse manifestations of the grieving process are being identified by a number of scholars and writers, on whom we can draw. Some of these are vernacular and emergent; others have been explicitly researched. Cunsolo Willox reported '*anticipatory* grieving for losses expected to come, but not yet arrived' (2012, p. 140, emphasis in original) among Canadian Inuit communities. Note that more than one author cited in this chapter and the next has connected their personal experiences of loss to their work (Zournazi 2002, Yusoff 2012, Hobbs 2013). There are resources in these writings for a wider societal learning process.

Loss of loved places

Scholars have identified elements of grief in the loss or change of loved places, as climate change transforms the places themselves, or disrupts the patterns of social life that interact with them (Hastrup 2009, Barnett & Campbell 2010, Farbotko & McGregor 2010, Cunsolo Willox 2012, Adger et al. 2013, Drew 2013). Most of these studies have, with good reason, focused on the implications for indigenous and developing world communities.

In interviews with Sami reindeer herders, Furberg et al. (2011, p. 8424) identified 'a sense of grief for the future'. Arctic latitudes have been among the first to experience significant climate change, including instability of seasonal changes that communities previously used to time activities associated with reindeer migrations. Considerable uncertainty was reported by study participants; but 'the herders did not consider *climate change* in itself to be the major threat to reindeer herding but rather to be *yet another stressor* on an already heavily burdened industry and culture' (Furberg et al. 2011, p. 8424, emphasis in original). Here they were referring to the multiple pressures on the herding lifestyle, many of which were felt to be beyond their control. Overall then the authors interpreted this community as 'facing the limit of resilience' (p. 8425). Attending to the importance of emotional health and wellbeing in responses of indigenous people to climate change also alerts us to the importance of these issues more broadly (Cunsolo Willox et al. 2013). The changes for those communities still living many aspects of a traditional lifestyle show a very direct connection (for example warmer winters constrain movements due to lack of safe access across ice, leading to reduced hunting possibilities and people being cooped up together for long periods of time), but the emotional implications for more urbanised communities may be just as strong, in different ways.

Albrecht et al. distinguished between loss of loved places (that they called nostalgia), and *solastalgia*: 'the pain or distress caused by the loss of, or inability to derive, solace connected to the negatively perceived state of one's home environment. Solastalgia exists when there is the lived experience of the physical desolation of home' (2007, p. 6). The environmental changes occasioning solastalgia are not only, or not necessarily, associated with climate change. Albrecht et al. used two examples; the effect of coal mining in Australia's Hunter Valley, and the influence of drought in a rural community in central western New South Wales. The distinction between nostalgia and solastalgia is an important one; we have never been able to hold places or environments as unchanging, but solastalgia acknowledges a different scale of loss. In relation to the health of the Murray River, Weir (2009, p. 134) discussed the importance of 'acknowledging ecocide', a challenge I return to below.

Non-human loss

The scale of species extinctions occurring as part of the Anthropocene has been the subject of considerable literature (van Dooren 2014), and is discussed further in Chapter 3. We are increasingly aware that the disappearance of charismatic fauna is only the tip of an iceberg. More humble species, many unknown or barely noticed in human lives, are also endangered and disappearing. We are yet to confront adequately the consequences of some of these for human wellbeing; Mathews (2011), for example, has written a poignant eulogy for the honeybee.

Complex questions arise here around whose lives are grieved for, and how (Cunsolo Willox 2012). Why do we worry more about the species than the individual as the unit of grief? As Ahmed reminds us, the politics of grief constitutes 'some others as the legitimate objects of emotion. This differentiation is crucial in politics as it works to secure a distinction between legitimate and illegitimate lives' (2004, p. 191).

Van Dooren's discussion of grief as a complex biosocial process that extends beyond the human provides an important contribution. He examines mourning among crows as a way to contest human exceptionalism:

Death does important boundary work in this kind of philosophical thought. Knowledge of death, or a relationship with the dead, here joins a long list of other "lacks", other characteristics or attributes that are thought to ground an essential difference between humanity and animality: be it the possession of language, mirror self-recognition, rationality, moral agency, or any number of other characteristics.

(2014, p. 132)

Grieving among elephants is also well known. Reviewing evidence from evolutionary psychology and ethology, van Dooren notes that grief 'is intimately entangled with the evolution of close social relationships', being part of the 'cost of commitment' for such relationships (2014, p. 134). Quite what we do with stories of crows mourning one another is an interesting question. For van Dooren, the act of bearing witness to loss is sufficient, or at least of value, in itself.

Yusoff extends this insistence to the question of violence – the violence involved in care and conservation – and its possibility as a space of potential: 'What if this violence was made available as a serious relation to think with alongside other entanglements with non-humans? And what if this violence, rather than being rendered invisible, was somewhere to meet?' (2012, p. 590). She takes up the point about how violence might be useful, although it is rarely considered as an imaginative resource: 'Why is it that we trust in the efficacy of beauty (non-human charisma, animal magnetism, vibrant matter, the 'wild' poetic moment), but do not yet want to trust in the efficacy of violence?' (2012, p. 587). These are important considerations when violence continues all around us, including towards the Earth itself.

Trauma of extreme events

Extreme events such as volcanic eruptions, earthquakes, bushfires and floods have been experienced throughout human history. Climate change projections are regionally variable but include increased frequency and/or intensity of many extreme events; floods, fires, cyclones. There is much to be learned from the increasing scholarly attention to the emotional dimensions of disaster response, as experienced from a number of different angles. Emotional outcomes are often much less visible than the more obvious losses of lives, homes and infrastructure. In the case of bushfires, for example, Team and Manderson show that they 'are characterized by high morbidity: burns, respiratory illnesses, psychological problems associated with fear, anxiety, grief, and loss among survivors and fire-fighters, and vicarious posttraumatic stress in the wider population' (2011, p. 906). Whittle et al. (2012) use the concepts of emotional work and emotional labour in disaster recovery, and their key findings resonate with thinking more broadly about grief in the Anthropocene:

- Much of the work of disaster recovery is hidden.
- It is not immediately obvious who is affected by a disaster.

- The emotional work of recovery can generate new vulnerabilities.
- Disaster recovery operates over timescales that are longer than expected.

The hidden aspects of negative or difficult emotions remind us to be alert for their expression, or non-expression, in different contexts. On the other hand trauma and loss, for example among bushfire victims, can also have positive outcomes, through enhancing senses of empowerment and belonging (Eriksen 2014). Increasingly, researchers who work with communities experiencing or recovering from disasters feel compelled to write movingly about their own emotional challenges in this process. This is often after a period of silence when they have felt that they should distance themselves from emotions in order to maintain their role as objective researchers (Dominey-Howes 2015), and/or because they do not want to suggest that their own trauma is equivalent to the more profound challenges being experienced in the researched community (Dominey-Howes 2015, Klocker 2015). Both Dominey-Howes and Klocker elaborate on complex layers of emotions, including guilt and powerlessness, that accompany the research process into traumatic events. Klocker analyses her own distress at 'not making a difference' to the lives of child domestic workers in Tanzania, noting that 'when conducting research on a traumatic issue, there is a great deal of pressure to achieve *something*' (2015, p. 5, emphasis in original). All writers in this emergent field discuss the importance of open and honest conversations about the process, without necessarily having 'answers' or 'solutions'.

Anxiety, denial and climate change

Moving to consider climate change specifically, perhaps most profound and widespread is the dimension of grief that manifests as denial. This is not confined to the so-called deniers and sceptics who have emerged so strongly in a number of Western countries, including Australia (Oreskes & Conway 2012). Even those who accept the science and have strongly supported various emissions reduction schemes have mostly imagined the necessary changes as a variation on business as usual rather than a more fundamental change. It is relevant here to explore the sense in which we are all climate change deniers.[4]

Weintrobe's (2013) application of psychoanalytic perspectives to anxiety around climate change provides important insights here, and is worth exploring in some detail. She discusses the subconscious conflict between two parts of our selves, a 'narcissistic part that hates reality, feels ideal and special and is prone to omnipotent thinking', and a realistic part that:

> tolerates limits, tolerates having very ambivalent feelings about reality, tolerates being far from perfect ... [and] is able to mourn an idealized world ... The narcissistic part is anxious it will not survive if reality is accepted. The realistic part is anxious that the narcissistic part has caused damage and may imperil its survival.
>
> (2013, pp. 33–34)

Common defences against these anxieties include 'feeling magically big and powerful' (p. 36) and denial, which can take two forms, negation and disavowal. The former is 'more likely to be a stage on the way to mourning illusions and accepting reality, [while] disavowal can involve the more stuck terrain of delusion' (p. 36).

'With disavowal, anxiety may be systematically gotten rid of, sometimes in a flash, through a range of "quick fixes". A central quick fix is minimizing or obliterating any sense that facing reality entails facing any loss' (2013, p. 39, see also Randall 2009).

These are not only experienced at the individual level but can be broader cultures of denial and disavowal, as I discuss below.

Specific anxieties about climate change include the following three:

- 'the loss of the Earth as the dependable bedrock that enables and supports our very life. . . . Specifically, we face the effects of a climate tipped into instability.'
- 'the survival of our very sense of self . . . our hope that we are generative – that our children will have children who will have children into the future – and rooted within long time.'
- Anxiety 'that our leaders are not looking after us'.

(Weintrobe 2013, pp. 42–43)

Contemplation of necessary lifestyle changes arouses anxieties in the narcissistic part of ourselves, since 'our identities and status are intimately bound up with our lifestyles' (p. 43). Weintrobe argues:

> I think in these circumstances what we dread giving up is not so much particular material possessions or particular ways of life, but our way of seeing ourselves as special, and as entitled, not only to our possessions but to our 'quick fixes' to the problems of reality. This underlying attitude, just one side of human nature, is strongly ingrained in current Western societies.
>
> (2013, p. 43)

There is considerable congruence between the conclusions of Randall (2009) and Weintrobe. It is important to acknowledge and speak our grief and loss; we need to provide emotional support for one another. 'It is important for people to bear their anxieties, because when they do not, their thinking deteriorates, and irrationality, lack of proportionality, hatred and narcissism are more likely to prevail' (Weintrobe 2013, p. 46). For Randall, it is only when we grieve that we will be able 'to remake our futures using all of our creativity, reason, feeling, and strength' (2009, p. 128).

We are all climate change deniers

In her ethnography of the pseudonymous Norwegian town of Bygdaby, Kari Norgaard (2011) has taken us on an important step from denial as a matter of individual psychology, to examine how it is socially organised and reproduced. The study throws light on the specific ways we proceed through everyday life as if things were otherwise. In so doing Norgaard adds to the widespread critique of the 'information deficit model' (Bulkeley 2000, p. 313) that implies better information and understanding of the science of climate change will be sufficient to solve its challenges.

Norgaard lived in the town during a winter when it was quite clear that something unusual was going on. The snows came late, and chantarelle mushrooms – a much loved part of the Scandinavian autumn – were available until the end of November. The disruption to the normal weather patterns of winter also disrupted the expected temporalities of social life. Skiing, an important part of local life, was delayed. Serving chantarelles for an Advent Sunday dinner was reported in the local paper as a clear disjuncture in social life. She asked how it is that a well-educated, politically active community, whose national identity coheres around ideals of environmental connection and egalitarianism, could keep climate change at a distance from their everyday lives? In contrast to the USA, where levels of actual climate science denial are higher, the Norwegians were not in dispute with the facts of climate change; they did not contest the science.[5] These are lives that had benefited over the last few decades from the proceeds of North Sea oil, thus benefiting from activities that had contributed to climate change.

Norgaard coined the term 'double reality':

> to describe the disjuncture I observed that winter in Bygdaby. In one reality was the collectively constructed sense of normal everyday life. In the other reality existed the troubling knowledge of increasing automobile use, polar ice caps melting, and the predictions of future weather scenarios.
>
> (2011, p. 5)

Drawing on work by Cohen (2001) and Zerubavel (2006) she built 'a model of socially organized denial . . . [which] emphasizes that ignoring *occurs in response to social circumstances* and *is carried out through a process of social interaction*' (Norgaard 2011, p. 9. emphasis in original). The social process took place through norms of conversation and norms of emotion. For example, it was acceptable to raise climate change in small talk at the bus stop, where people might comment on the strangeness of the weather, but not in educational settings or conversations between parents and children, where people felt pressure to be optimistic rather than scary. Emotional norms, especially for men, involved being strong. Emotional responses to climate change were thus little discussed, except in ways that invoked humour or irony, or in some social gatherings where alcohol was involved. The lack of conversational norms around climate change was also identified as significant by Rowson (2013), who found that only 60 per cent of a UK study sample had ever spoken about climate change, and the majority of those for less than ten minutes. He interpreted these as conversations that are cut short when they become uncomfortable, highlighting 'that there is no meaningful national conversation about climate change' (p. 8). Individual acts were characterised by Norgaard as 'turning away', partly so as not to get depressed, as a number of her younger informants expressed it.

These individual acts were supported by wider cultural narratives, especially around the nation. 'By portraying Norwegians as close to nature, egalitarian, simple, and humble, these narratives of national identity counter the criticism that Norwegians face with regard to climate change and petroleum policies' (Norgaard 2011, p. 140). In these and many other examples, Norgaard documents the way 'it *takes work* to ignore the proverbial elephant in the room' (p. 93, emphasis added). Cultural norms and narratives are used to collectively hold climate change at arm's length; 'public nonresponse to global warming is *produced* through cultural practices of everyday life' (p. 207, emphasis in original). In a smaller-scale study, Norgaard makes comparisons with everyday denial in the USA, arguing that it is similar, and likely to be reflect wider trends across the affluent West. 'It is significant that what I describe as "climate denial" felt to people in Bygdaby (and, indeed, to people around the world) like "everyday life"' (p. 121).

Although everyday practices reproduce traditions and expectations, in Bygdaby more than many other places, they can also be done differently. One hopeful implication of this research, chiming with issues and experiences explored in the second half of this book, is to consider ways in which the everyday can be reconfigured differently, contributing to necessary social change from the ground up.

Rowson (2013) identified a similar phenomenon, that he called 'stealth denial', in about two thirds of the British public. His survey results showed that while only a fifth of the population would be termed deniers or sceptics, two thirds:

> whose views are generally considered unproblematic are 'unmoved' in the sense that they do not accept the full implications in terms of their feelings, agency and complicity. This group corresponds to those who accept the reality of anthropogenic climate change, but gave answers suggesting they didn't appear to have the commensurate feelings, sense of responsibility or agency that one might expect. Only a small group (14.5 percent) seem to live in ways that are relatively consistent with their understanding of the problem.
> (2013, p. 7)

Considerable research effort has gone into understanding the cultural basis of climate change denial among the 20 per cent. This denial splits from acceptance of climate change science along a number of divides; conservative rather than liberal political orientation, males rather than females, older rather than younger generations and so-called materialist rather than postmaterialist values (Dunlap & McCright 2008, Tranter 2011, Tranter 2014). These cultural and psychological analyses have explored complex and intersecting influences on the constitution of denial, for example the independent roles of morality around maintaining the social order, and that supporting economic liberty (Rossen et al. 2015), and the concern with purity and sanctity (Feinberg & Willer 2013).

Less explicit research effort has gone into understanding Rowson's two thirds of stealth deniers, or the majority groups who accept the scientific facts but are still 'living in denial', to use Norgaard's terminology, but there is a growing body of work that throws light on these apparently contradictory environmental cultures. The surveys from which statistical results can be derived are reliable in terms of attitude and cognition, and contribute to problems such as denial being framed in those terms. Research in the ethnographic tradition, such as Norgaard's important contribution, shows how much more complex actual practice is.[6] A large body of work, including that discussed in Chapter 8, now shows considerable discrepancy between attitude and practice in relation to environmental issues. A notable example is in the extent to which people who espouse strong environmental values continue to undertake air travel (Barr et al. 2010; Barr et al. 2011; Waitt et al. 2012). There are multiple reasons for these contradictions and complexities, and some of them are discussed in the second half of the book. They include the influence of embedded infrastructures, and socioeconomic structures that limit effective choices for many people. Rowson (2013), for example, sees two obstacles as closely related. First, the focus on deniers and sceptics is 'a radical misunderstanding of the nature and ubiquity of climate change denial as something purely cognitive' (pp. 5–6). Second, in the UK there is:

> a significant underestimation of rebound effects on emissions reductions made through energy efficiency. These two elements of the issue are rarely placed together, but appear to be closely connected; we place hope in relatively ineffectual actions because we haven't fully faced up to the nature of the problem.
>
> (2013, p. 6)

I return to the connections between denial and action, or lack thereof, in Chapters 6 and 8. Here I want to hold the focus on the emotional dimensions of grief and denial.

Norgaard puts emotions and emotional norms at the centre of her analysis. She emphasises that while denial has a negative connotation:

> nonresponse is not a question of greed, inhumanity, or lack of intelligence. Indeed, if we see information on climate change as being *too disturbing* to be fully absorbed or integrated into daily life ... this interpretation is the very opposite of the view that nonresponse stems from inhumanity or greed. Instead, denial can – and I believe should – be understood as testament to our human capacity for empathy, compassion, and an underlying sense of moral imperative to respond, even as we fail to do so.
>
> (2011, p. 61)

These are challenging, and charitable, thoughts that clearly contest one of the Anthropocene narratives, that humanity is an inherently selfish and destructive species. Norgaard's argument that we block out these thoughts *because* we care so much helps us find a pathway forward. Bringing elements of the arguments of Weintrobe, Randall and Norgaard together, we have an argument that we keep climate change at arm's length to protect parts of our identity. This takes cultural work, and change will take cultural work as we transition – by force or choice – to new identities and practices. Part of that cultural work is grieving for what we might lose in the process.

Grief, modernity and the modern self

The grieving for modernity that I have been describing is amorphous. It has diverse manifestations, and is experienced differently by different people. But it shares the idea of a hopeful future, built on the Enlightenment ideal of improvement which 'transmitted a belief in the possibilities of progress into the fabric of everyday life' (Gascoigne 2002, p. 10). That such aspirations continue to be shared by those on different ends of the political spectrum is illustrated in Australian politics, with:

> the Left looking to the prospects for social betterment offered by a state which serves its citizens through careful planning and the cultivation of improving institutions such as schools, gaols and hospitals, the Right to the widening horizon of economic growth which the programme of Adam Smith had laid down in the language of the Scottish Enlightenment.
> (Gascoigne 2002, p. 169)

If the modern self is not a single phenomenon, then neither will be its loss. Nor is the modern self – despite its claims to individual freedom – a free-standing entity. We are entangled in various ways and will need to grieve those entanglements; 'Loss requires mourning and grieving for the destruction of a relation and those subjects that are constituted through that relation' (Yusoff 2012, p. 579).

Climate change challenges a number of aspects of the modern self, including the way we are all embedded in the fossil fuel economy that must be undone. Weintrobe (2013) argued that we are in the process of losing that part of our sense of identity connected to the material possessions of a consumerist society. While that may be true, my feeling is that radical change in the availability of material possessions will be the least of our problems, and will even be welcomed by some. The modern dream of wealth – 'the twentieth century [that] looked like it was here for good' (Urry 2011, p. 48) – is so contingent and short-lived in the context of human history that we have many capacities to do otherwise, as many societies across the world continue to demonstrate. Rather, the modern subject I see as becoming radically undone has more in common with Hamilton, who argued that: 'The foundational beliefs of modernity – the unlimited scope of human achievement, our capacity to control the world around us, our belief in the power of knowledge to solve whatever discomforts us – will collapse' (2010, p. 210).

Those foundational beliefs have always contained many contradictions. We have not all shared in human achievements equally; the improved wellbeing through modern healthcare systems has itself been embedded in and dependent on a fossil fuel economy; the control we believed we had was more of a temporal displacement of the effects of our actions, which are now coming back to bite us; and so on. The 'need to believe that the world is just, orderly, and stable, a motive that is widely held and deeply ingrained in many people' (Feinberg & Willer 2011, p. 34) contributes to our collective denial.

The modern self is tempted to move quickly to solutions, to try and work out what we need to do to 'fix' things. I discuss some of this in the second half of the book, but it is more important first to acknowledge and name our grief, sit quietly with it and, as Klocker (2015) writes, to experience the distress of not making a difference. There are parallels here with Weir's take on changes along the Murray River. She writes, 'There is something we must do before we can begin to do anything about river destruction. We have to look directly at "ecocide" and acknowledge that it has happened' (2009, p. 134).

For Weir, modern water managers must ignore the consequences of their work ('the desiccation of river life, the loss of river variability, the rising water-table, and so on') because 'to hear this message is to unsettle the foundations of their management approach' (2009, p. 134). In contrast, traditional owners can 'acknowledge the loss of river life without unsettling the foundations of their knowledge' (p. 135). The evidence from the psychological and psychoanalytical literature is clear: the first step is to acknowledge this companion and to talk about it more openly and honestly. Weintrobe (2013) does not define what she means by 'bearing' our anxieties, and it has the self-evident meaning of carry. But the tone of these arguments is broader, perhaps in the mode of child-bearing; it also about holding, articulating, bearing witness and rendering visible.

If denial is part of grief, and grief will be our companion, it follows that denial will also be our companion. Denial and grief will be wrapped up with our hopeful practices (Chapter 5). For other psychological reasons that are probably separate from grief, our denial is not going to go away either. As Rowson says:

> The very notion of denial, in which we somehow simultaneously know something and yet choose not to face up to that knowledge – is perplexing when the working assumption is that human beings are unitary, rational and self-consistent. However, denial begins to look normal, even adaptive, when you realise that our sense of self is constructed from a coalition of fragments, that most of what we do is unconscious, that we are motivated to keep feeling good about ourselves, and that we are, in many ways, strangers to ourselves.
> (2013, p. 36)

There are particular responsibilities for scholars in this kind of bearing, and there are also risks. The work we have to do is emotional as well as everything else (Whittle et al. 2012; Robbins & Moore 2013). Grief is our companion in the research enterprise as well as in everyday life. As researchers, we are not immune from the emotional costs of this, and need to look out for one another. This is discussed further in Chapter 5.

The productive potential of mourning

I am not claiming that acknowledgement of grief will help those of us embedded in fossil consumption throw fossil capital out of the driver's seat, to use Malm's (2014, p. 743) metaphor. However I do argue that we will not be able to make the transition to low carbon societies without it. On the psychoanalytical and psychological readings discussed above, grief is not something that we can get 'beyond', rather it has to become part of our lives and politics. But there is much to be gained by reflecting on Randall's argument that: 'A more sophisticated understanding of the processes of loss and mourning, which allowed them to be restored to public narratives, would help to release energy for realistic and lasting programmes of change' (2009, p. 118). In her analysis, there is work involved; 'the work of grief is a series of tasks that can be embraced or refused, tackled, or abandoned . . . [the work is] always in progress, never complete' (p. 121). The tasks include accepting the reality of loss, first intellectually then emotionally; working through the painful emotions of grief; and adjusting to the new environment/acquiring new skills/developing a new sense of self (Randall 2009, Table 1). Only then can emotional energy be reinvested in more creative ways. Bearing our grief will not necessarily stave off catastrophe, but it will give us a better chance of effective action.

The specific expressions of grief will be variable and much more discussion is needed as to how it can be incorporated more creatively into everyday life. Cunsolo Willox (2012, p. 137) has started this conversation by conceptualising climate change as 'the work of mourning', and considering the potential for such labour to be productive. Drawing on Butler (2004), she argues:

> grief and mourning have the unique potential to expand and transform the discursive spaces around climate change to include not only the lives of people who are grieving because of the changes, but also to value what is being altered, degraded, and harmed as something mournable.
>
> (2012, p. 141)

As many public mourning activities – focused on war, disaster or disease, for example – have shown, collective grieving can bring people together, provide comfort, expose relational ties and create a 'we-ness'. Those relational ties include humble recognition of the many non-humans on whom our own lives are dependent. In so doing it can create potent opportunities for transformation (Cunsolo Willox 2012). 'Environmentally-based grief needs to continue to be spoken loudly and often, in private and public settings' (p. 151); here she sees a particular role for art, literature and music in speaking the unspeakable. Other scholars who have also come to this point include Evans and Reid (2014), and the Dark Mountain Project's commitment to 'uncivilised writing' (Kingsnorth & Hine 2009).

If part of what we are grieving for, and what we must farewell, is our modern selves, it follows that a necessary intellectual and practical task is to imagine new kinds of selves. What range of characteristics will contribute to low carbon subjectivities? What relations will constitute an Anthropocene self, and how much will this have in common with the Moderns? Returning to consider again the grieving homeowners of the Blue Mountains bushfires, we can wonder whether they are already Anthropocene subjects; living with uncertainty, relinquishing household insurance, living with lower levels of material 'stuff'. We return to the specifics of these processes in the second half of the book. In Chapter 3 we examine how these issues play out in a field where grief has already been an explicit part of the conversation around environmental change – how it has infused environmental thought.

Notes

1 www.3aw.com.au/blogs/neil-mitchell-blog/adam-bandts-indecent-insensitive-bushfires-tweet/20131018-2vqgk.html. Last accessed November 2013.
2 www.uowblogs.com/ausccer/2013/10/23/reflections-from-the-fire-front-and-research-in-its-ashen-wake/. Last accessed August 2015.
3 In response to Noam Chomsky's column title 'The End of History' (in.is/ow.ly/K3yBU) Luciano Floridi (@Floridi) tweeted, 'IPCC report is reliable and very worrying, but pessimism's bitter pleasure is unhelpful'.
4 Rowson (2013, p. 36) quotes Hamilton (2013), as saying 'We are all climate deniers'.
5 This may be changing; a recent study indicates that levels of climate change denial are as high in Norway as in Australia, New Zealand and the USA (Tranter & Booth 2015). Tranter and Booth point out that these are all countries with high per capita greenhouse gas emissions.
6 Jackson also uses the concept of denial in his discussion of how to decouple prosperity from growth:

> The starting place must be to unravel the forces that keep us in damaging denial. Nature and structure conspire together here. The profit motive stimulates a continual search for newer, better or cheaper products and services. Our own relentless search for novelty and social status locks us into an iron cage of consumerism. Affluence itself has betrayed us.
>
> (2009, p. 188)

References

Adger, W. N., J. Barnett, K. Brown, N. Marshall and K. O'Brien. 2013. "Cultural dimensions of climate change impacts and adaptation." *Nature Climate Change* 3(2): 112–117.
Ahmed, S. 2004. *The Cultural Politics of Emotion*. New York: Routledge.
Albrecht, G., G.-M. Sartore, L. Connor, N. Higginbotham, S. Freeman, B. Kelly, H. Stain, A. Tonna and G. Pollard. 2007. "Solastalgia: The distress caused by environmental change." *Australasian Psychiatry* 15: S95–S98.
Barnett, J. and J. Campbell. 2010. *Climate Change and Small Island States: Power, Knowledge, and the South Pacific*. London: Earthscan.
Barr, S., G. Shaw and T. Coles. 2011. "Times for (un) sustainability? Challenges and opportunities for developing behaviour change policy. A case-study of consumers at home and away." *Global Environmental Change* 21(4): 1234–1244.

Barr, S., G. Shaw, T. Coles and J. Prillwitz. 2010. "'A holiday is a holiday': Practicing sustainability, home and away." *Journal of Transport Geography* 18(3): 474–481.
Bondi, L., 2005. "Making connections and thinking through emotions: Between geography and psychotherapy." *Transactions of the Institute of British Geographers* 30: 433–448.
Bulkeley, H. 2000. "Common knowledge? Public understanding of climate change in Newcastle, Australia." *Public Understanding of Science* 9(3): 313–333.
Butler, J. 2004. *Precarious Life: The Powers of Mourning and Violence.* London: Verso.
Cohen, S. 2001. *States of Denial: Knowing About Atrocities and Suffering.* Cambridge: Polity Press.
Cook, B. R. and A. Balayannis. 2015. "Co-producing (a fearful) Anthropocene." *Geographical Research* 53: 270–279.
Cunsolo Willox, A. 2012. "Climate change as the work of mourning." *Ethics and the Environment* 17(2): 137–164.
Cunsolo Willox, A., S. L. Harper, V. L. Edge, K. Landman, K. Houle, J. D. Ford and the Rigolet Inuit Community Government. 2013. "The land enriches the soul: On climatic and environmental change, affect, and emotional health and well-being in Rigolet, Nunatsiavut, Canada." *Emotion, Space and Society* 6: 14–24.
Dominey-Howes, D. 2015. "Seeing 'the dark passenger' – Reflections on the emotional trauma of conducting post-disaster research." *Emotion, Space and Society.*
Drew, G. 2013. "Why wouldn't we cry? Love and loss along a river in decline." *Emotion, Space and Society* 6: 25–32.
Dunlap, R. and A. McCright. 2008. "A widening gap: Republican and democratic views on climate change." *Environment* 50(5): 26–35.
Eriksen, C. 2014. *Gender and Wildfire: Landscapes of Uncertainty.* New York: Routledge.
Evans, B. and J. Reid. 2014. *Resilient Life: The Art of Living Dangerously.* Cambridge: Polity Press.
Farbotko, C. and H. McGregor. 2010 "Copenhagen, climate science and the emotional geographies of climate change." *Australian Geographer* 41(2): 159–166.
Feinberg, M. and R. Willer. 2011. "Apocalypse soon? Dire messages reduce belief in global warming by contradicting just-world beliefs." *Psychological Science* 22(1): 34–38.
Feinberg, M. and R. Willer. 2013. "The moral roots of environmental attitudes." *Psychological Science* 24(1): 56–62.
Furberg, M., B. Evengård and M. Nilsson. 2011. "Facing the limit of resilience: Perceptions of climate change among reindeer herding Sami in Sweden." *Global Health Action* 4: 8417–8427.
Gascoigne, J. 2002. *The Enlightenment and the Origins of European Australia.* Cambridge: Cambridge University Press.
Hamilton, C. 2010. *Requiem for a Species: Why We Resist the Truth About Climate Change.* Sydney: Allen and Unwin.
Hamilton, C. 2013. "What history can teach us about climate change denial." In *Engaging with Climate Change: Psychoanalytic and Interdisciplinary Perspectives*, edited by S. Weintrobe, 16–32. Sussex: Routledge.
Hastrup, K. 2009. "The nomadic landscape: People in a changing Arctic environment." *Geografisk Tidsskrift-Danish Journal of Geography* 109(2): 181–189.
Hobbs, R. J. 2013. "Grieving for the past and hoping for the future: Balancing polarizing perspectives in conservation and restoration." *Restoration Ecology* 21(2): 145–148.
Jackson, T. 2009. *Prosperity Without Growth: Economics for a Finite Planet.* London: Earthscan.

Kingsnorth, P. and D. Hine. 2009. "Uncivilisation: The Dark Mountain Manifesto." The Dark Mountain Project. Accessed 23 July 2015. Available at http://dark-mountain.net/about/manifesto/.

Klocker, N. 2015. "Participatory action research: The distress of (not) making a difference." *Emotion, Space and Society*. DOI:10.1016/j.emospa.2015.06.006.

Koskela, H., 1997. " 'Bold walk and breakings': Women's spatial confidence versus fear of violence." *Gender, Place and Culture* 4: 301–319.

Malm, A. 2014. *Fossil Capital: The Rise of Steam-Power in the British Cotton Industry, c. 1825–1848, and the Roots of Global Warming*, PhD Thesis. Lund: Lund University.

Mathews, F. 2011. "Planet Beehive." *Australian Humanities Review* 50: 159–178.

McKinnon, C., 2014. "Climate Change: Against Despair." *Ethics and the Environment* 19: 31–48.

Norgaard, K. M. 2011. *Living in Denial: Climate Change, Emotions, and Everyday Life*. Cambridge, MA: MIT Press.

O'Neill, S. and S. Nicholson-Cole. 2009. "'Fear won't do it': Promoting positive engagement with climate change through visual and iconic representations." *Science Communication* 30: 355–379.

Oreskes, N. and E. M. Conway. 2012. *Merchants of Doubt*. New York: Bloomsbury.

Pahl, S., S. Sheppard, C. Boomsma and C. Groves. 2014. "Perceptions of time in relation to climate change." *WIREs Climate Change* 5: 375–388.

Randall, R. 2009. "Loss and climate change: The cost of parallel narratives." *Ecopsychology* 1(3): 118–129.

Rickards, L., J. Wiseman and Y. Kashima. 2014. "Barriers to effective climate change mitigation: The case of senior government and business decision makers." *WIREs Climate Change* 5: 753–773.

Robbins, P. and S. A. Moore. 2013. "Ecological anxiety disorder: Diagnosing the politics of the Anthropocene." *Cultural Geographies* 20(1): 3–19.

Roeser, S. 2012. "Risk communication, public engagement, and climate change: A role for emotions." *Risk Analysis* 32: 1033–1040.

Rose, D. B. 2013. "Anthropocene Noir." *Arena Journal* 41/42: 206–219.

Rossen, I. L., P. D. Dunlop and C. M. Lawrence. 2015. "The desire to maintain the social order and the right to economic freedom: Two distinct moral pathways to climate change scepticism." *Journal of Environmental Psychology* 42: 42–47.

Rowson, J. 2013. *A New Agenda on Climate Change: Facing Up to Stealth Denial and Winding Down on Fossil Fuels*. London: Royal Society for the Encouragement of Arts, Manufactures and Commerce.

Team, V. and L. Manderson. 2011. "Social and public health effects of climate change in the '40 South'." *WIREs Climate Change* 2(6): 902–918.

Tranter, B. 2011. "Political divisions over climate change and environmental issues in Australia." *Environmental Politics* 20(1): 78–96.

Tranter, B. 2014. "Social and political influences on environmentalism in Australia." *Journal of Sociology* 50(3): 331–348.

Tranter, B. and K. Booth. 2015. "Scepticism in a changing climate: A cross-national study." *Global Environmental Change* 33: 154–164.

Urry, J. 2011. *Climate Change and Society*. UK: Polity.

van Dooren, T. 2014. *Flight Ways: Life and Loss at the Edge of Extinction*. New York: Columbia University Press.

Waitt, G., P. Caputi, C. Gibson, C. Farbotko, L. Head, N. Gill and E. Stanes. 2012. "Sustainable household capability: Which households are doing the work of environmental sustainability?" *Australian Geographer* 43(1): 51–74.

Weintrobe, S. 2013. "The difficult problem of anxiety in thinking about climate change." In *Engaging with Climate Change: Psychoanalytic and Interdisciplinary Perspectives*. edited by S. Weintrobe, 33–47. Sussex: Routledge.

Weir, J. K. 2009. *Murray River Country: An Ecological Dialogue with Traditional Owners*. Canberra: Aboriginal Studies Press.

Whittle, R., M. Walker, W. Medd and M. Mort. 2012. "Flood of emotions: Emotional work and long-term disaster recovery." *Emotion, Space and Society* 5(1): 60–69.

Yusoff, K. 2012. "Aesthetics of loss: Biodiversity, banal violence and biotic subjects." *Transactions of the Institute of British Geographers* 37(4): 578–592.

Zerubavel, E. 2006. *The Elephant in the Room: Silence and Denial in Everyday Life*. Oxford: Oxford University Press.

Zournazi, M. 2002. *Hope: New Philosophies for Change*. Annandale, Australia: Pluto Press.

3 Past, present and future temporalities

> *The photographs in this section are of a lake which no longer exists. It lies fifty feet beneath the surface of an impoundment of water twenty-five times its size, and still bearing its name.*
>
> *The flooding of the vast glacial valley in which Lake Pedder rested occurred in 1972. The water thus collected was designed to add to the main storage in the Gordon river valley as part of a hydro-electric power scheme.*
>
> *One important feature of the lake was its magnificent beach, formed by countless centuries of westerly winds and wave action. The beach (in reality the floor of the lake), was covered during the wet winter months to the foot of the sand dunes which formed its eastern boundary.*
>
> *Each summer, as the water level dropped, the bed of the lake was revealed as a "beach" of such magnitude that it could have comfortably supported the City of Sydney on its level surface. It was two miles in length, and reached up to eight hundred yards in width in mid-summer.*
>
> *As a self-renewing natural landing ground for aircraft it was beyond price. As a work of nature its haunting beauty was beyond compare. The fourteen plates in this section recall the vanished splendour of this once unique place.*
>
> (Angus 1975, p. 88)

There is no more potent symbol of environmental grief in Australia than Lake Pedder (Figure 3.1). The words above form the introduction to a special Lake Pedder section of Max Angus's book in memory of the acclaimed wilderness photographer and environmental activist Olegas Truchanas. Truchanas had used a slide show[1] of Lake Pedder, set to music, as part of the public campaign to save the lake. The loss of Lake Pedder galvanised a set of Tasmanian environmental campaigns throughout the latter decades of the twentieth century. Indeed the rise of Green politics, first in Tasmania and then nationally, can be traced to mourning as a result of this environmental destruction. Some people continue to work towards the lake's restoration.[2] Tasmania occupies a special place in the history of environmental thought and activism, illustrating both the importance and the pitfalls of mourning for lost landscapes.

This chapter connects the discussion of grief to the issue of temporality. I focus on the pitfalls of mourning; to the extent that it is expressed as a longing for a

Figure 3.1 Sand waves in the shallows of Lake Pedder with a view to the plains around Maria Lake, Tasmania, 1968. Photographer Olegas Truchanas. Digital copy courtesy National Library of Australia.

pristine past, it hampers our capacity to approach Anthropocene futures. My argument builds through several steps. First, I examine expressions of grief and loss in the biodiversity conservation literature, arguing that the 'Edenic Sciences', which include conservation biology, restoration ecology and invasion biology (Robbins & Moore 2013), have a particular issue with grief because of their adherence to a past baseline. Indeed that grief is exacerbated *because* we conceptualise a pristine, linear past. The concept of returning to baselines is not only historically inaccurate, it also entrenches an unduly linear understanding of temporality. Second, the timelines of the colonial imaginary provide a key example because of the way they constitute the temporal baseline of environmental management. Further, biodiversity conservation debates have not constituted their concepts of time out of thin air. They are part of the temporalities of modernity itself with its focus on progress and improvement through time.

Such a view of history has always been less accurate than one in which contingency and non-linear change was foregrounded, as discussed in the third section of the chapter. But it is particularly inappropriate for the Anthropocene, a defining characteristic of which is an accelerated scale of non-linear change and increased uncertainty. The Anthropocene is a concept hewn out of past, geological time, now being used to characterise a human future (or lack thereof). There may be debate about whether it started 8000 years ago, 300 years ago or 50 years ago, but it is always shown on a timeline. So my argument applies also to the emerging discourse of the Anthropocene; there is a risk that we reify it as just another stage along a timeline of history. If temporalities focused on the past have dominated recent Western environmental thinking, they have also shaped the spatial practices of environmental management and protection. In the final sections of the chapter we consider how more dynamic temporalities might take expression in different land management and biodiversity conservation decisions.

Nothing in this chapter should be read as belittling mourning for lost places, or as undoing the argument in the previous chapter about the importance of grieving. Rather it gives us pause to consider carefully what we are grieving for, and whether we ever actually had it. The myth of unpeopled wilderness has already taught us part of this lesson. Many of the images that Olegas Truchanas used in his slide show contained people, including some beautiful pictures of his own children. There were pictures of the planes that brought people to the beach to camp, and pictures of their tents. This is notably different from the images used in the wilderness campaigns of the 1980s and 1990s, none of which contained signs of any human presence. It is that later erasure of human presence that has been comprehensively critiqued (Cronon 1996).

It is also important throughout the chapter to expand our thinking about temporalities, drawing on scholars who have shown how 'social practices and natural processes are bound together in a relationship of co-construction, in which "consonance or dissonance" between the two is produced through different forms of timing, tempos, and rhythms' (Pahl et al. 2014, p. 377). Different ways of life are attuned to and construct different temporalities (Bastian 2012). For example, agricultural practices before the Industrial Revolution 'were linked strongly to natural rhythms of climate variation and seasonal growth and decay' (Pahl et al. 2014, p. 377). Fishers are linked to tidal rhythms. And seasonal calendars of indigenous people are based on relationships between different phenomena (Green et al. 2010, Lefale 2010), such as connection between the flowering of particular trees and the arrival of particular fish. The disruption of such relations is often the basis on which people discern how the climate is changing (Green & Raygorodetsky 2010). This is a huge field of literature, and I can examine only a few aspects of temporality: the role of linear baselines, the possibilities of contingent and uncertain futures and the everyday temporalities of climate change.

Grief, loss and the lure of the past baseline

In his farewell editorial after a decade at the helm of the journal *Restoration Ecology*, Richard Hobbs (2013) identified different patterns of grieving – between those clinging to the hope of a pristine past, and those accepting of a messier future – as one of the sources of disagreement in biodiversity conservation science. Hobbs talked of conservation biologists being constantly assailed by loss, and noted that the recognition of this has a long history (Windle 1992). This history continues as writers mourn for species, habitats and communities disappearing as part of what has been called the sixth mass extinction (Barnosky et al. 2012). Hobbs accounted for polarised views about how much to accept the state of ongoing change in ecosystems by appealing to the stages of grief popularised several decades ago by Elisabeth Kubler Ross.[3] He drew attention to the divisive nature of debates in conservation science, connecting anger and denial in scientific debate to equivalent emotions in the grieving process. Examples include debates over natives vs non-natives (Davis et al. 2011, Simberloff 2011) and more recent debates over novel ecosystems (Murcia et al. 2014, Hobbs et al. 2014, Aronson et al. 2014). Much of the critique of the so-called 'new conservation' paradigm, which accepts and tries to deal with the pervasiveness of human presence and the transformations of the Anthropocene, is couched in terms of whether such people have 'given up hope' for the future. (For a recent example of how heated these debates can become, see Miller et al. 2014 and Marris 2014).

While grief is now understood as a more complex process than the linear 'stages' that Kubler Ross identified, even to label it as an issue in the supposedly dispassionate sciences is an important step. Randall (2009, p. 121) argued that the main limitation of the Kubler Ross model for grief associated with climate change is that 'it is a model from the end of life. It describes an experience without transition and without hope'. In keeping with the overall argument of this book, it is important to find ways to carry our grief into hopeful environmental engagements. This means probing exactly what we are grieving for. A particular influence in conservation science and activism, it seems, is the influence of past conditions as providing the baseline. Against such an ideal, the present can hardly be understood in any other terms than loss.

Environmental restoration has within it the ideal of healing, but also implies a baseline of time, a time to which we are returning. These concepts – conservation, restoration, preservation – encapsulate the past within the structure of the words themselves. Hinchliffe (2008) described conservation as something that comes 'after' nature, the rationale being to return things to pre-existing states. Also in the British context, Lorimer (2008) detailed the lure of the Clementsian climax – one end point in an inherent cyclicity – as the temporal goal towards which management aspires. The lively debate around 'rewilding' also addresses these issues in relation to returning Pleistocene megafauna to the landscapes of Europe and North America (Lorimer & Driessen 2013).

So, how are the ecologists dealing with their grief? It is arguably not as well resolved as it could be. For Robbins and Moore (2013), the fervency of the international debates to which Hobbs refers, for example over alien species, is indicative of a larger upheaval they call Ecological Anxiety Disorder, within the Edenic Sciences. These sciences

> share a tacit epistemological commitment to evaluating ecological relationships explicitly with regard to an a priori baseline – a condition before the Columbian encounter, or a time or place before human contact, or a place of expulsion or return – one Before the Fall.
>
> (Robbins & Moore 2013, p. 4)

Hulme (2009) has also identified what he calls the 'Lamenting Eden' myth in relation to climate change – a vision of a natural, stable state of the Earth to which we can aspire and return. For Robbins and Moore, the anxiety in question is partly about the disintegration of the self, as evidence for the baseline recedes. They argue that part of the 'treatment' is to be more explicit about the political choices involved in environmental intervention, and admit the emotional and intellectual struggle that follows from embracing environmental novelty; thus 'we might conquer our phobias and dispense with imaginary places to which there is no hope of return' (Robbins & Moore 2013, p. 16). It is hard to escape the conclusion that the division of which Hobbs writes is between those who, in Weintrobe's (2013, p. 34) terms are 'able to mourn an idealized world', and those who are in a degree of denial about how much it has changed.

Timelines exacerbated by the colonial imaginary

The Edenic Sciences are particularly strong where there is a context of colonial history. There is more than can be said here about the conservation cultures of late modernity over the last fifty years, but it is important to remark on the close connection between invasion ecology and New World contexts such as North America, Australasia and South Africa. This is not only because of the historic spread and interchange of different and previously disjunct peoples, animals and plants (Crosby 2004), many of them damaging; it is also because of the temporal threshold that was crossed between non-history and history.

In colonial contexts the temporal lure is not only backwards, but also to a firm baseline of pristineness corresponding to the point before colonisation. Rose (1997) calls this Year Zero, marking the beginning of history. She places the Western co-location of hope and the future in the linear yet palinstrophic timing of the Bible, arguing that that time (Year Zero) provides the central mediating point, moving out from which there is an identical structure in each direction. Thus events in the Hebrew Bible prefigure events that are then fulfilled in the Christian Bible. Adam prefigures Jesus; the initial creation of the Earth prefigures the final destruction and creation of a new heaven and new earth. There is a point to history. 'The concept of an irreversible sequence shapes time, and has proved to be a powerful tool in modern culture' (Rose 1997, p. 27). Rose's main point in that chapter is to understand the connected spatial logic of the frontier and its implications for indigenous people. 'Ground Zero is also the Year Zero: the moment at which history will begin with the arrival of the outriders of civilisation' (Rose 1997, p. 26).

> I am contending that in settler societies the frontier is culturally constructed as precisely this moment: a disjunction between wholly different kinds of time. I suggest that we imagine the frontier as a rolling Year Zero that is carried across the land cutting an ontological swathe between 'timeless' land and historicised land.
>
> (Rose 1997, p. 28)

Just as Rose extended her argument to the way in which pastoralists and rangeland pastoralism now face an uncertain future, we can consider what it would mean to leave the comforts and apparent certainties of linear time behind.

The European colonisation of Australia marks such a Year Zero in the way it defines a temporal threshold of nativeness in our thinking and environmental practice. The year 1788 is assumed as the baseline of naturalness and/or nativeness in Australian endangered species legislation (Lien & Davison 2010), in government measurements of environmental health (Trigger 2008), in river management (Lavau 2011), in Commonwealth government 'Working on Country' programmes (Martin & Trigger 2015) and in key government overviews such as the State of Environment reports (Beeton et al. 2006). Or, for plants, the year is 1770, when Joseph Banks and Daniel Solander spent eight days collecting 132 plant species on the shores of Botany Bay in late April to early May 1770, 'spreading them upon a sail in the sun' (Benson & Eldershaw 2007, p. 118) so that their samples did not spoil. This was the first scientific collection of Australian flora, an event and a collection of immense historical and scientific significance. But did this event usher in a new ontological state? Did the iconic *Banksia serrata* pass into a fundamentally different state of being by being pressed, dried and transported to England? Of course not, but, as Lavau (2015) argues, this temporal threshold is generally considered to be so self-evident that it is not worth discussing. These issues are explored further in Chapter 7. Here we should note that the temporal threshold is also inescapably spatial, encapsulated in Rose's metaphor of the frontier as a rolling Year Zero.

From teleology to contingency

The colonial understanding of history was being shaped by 'that key Enlightenment precept: the possibilities of progress' (Gascoigne 2002, p. 10), and the seventeenth and eighteenth-century turn towards a grand, teleological story of human improvement (Anderson 2007). Anderson argued that the colonial enterprise did not simply fan pre-formed ideas across the globe, rather these ideas were constituted in the process of colonialism. Thus 'evolutionary theorising was not so much applied to Australia, as *constituted* out of materials from a time/space location where the archaic human first began lifting himself up' (Anderson 2007, p. 176, emphasis in original). The movement of civilisation was widely understood as forward through time, and upward out of nature. We return in Chapter 4 to the details of how 'human distinction became elaborated as a stadial movement out of nature' (Anderson 2007, p. 42), and the crucial role of agriculture in defining this move. Here we focus on the progressivist forward movement. Anderson argues that Enlightenment faith in human perfectibility, as seen, for example, in the work of John Locke, 'was underpinned by a series of (violent) imperial notions' (p. 46). The first of these was 'history as a march through time (beyond a state of nature)' (p. 46). She quotes Tylor on the character of:

> Civilization . . . as in personal figure she traverses the world . . . and if now and then she tries a few backward steps . . . it is not in her nature. Her feet

were not made to plant uncertain steps behind her, for both in her forward view and in her onward gait she is of truly human type.

(2007, p. 171 quotes Tylor 1958 (1871), p. 69)

This ideal of improvement applied to both land and people (Gascoigne 2002). It took particular root in Australia, where even the plants were found to be weak (Anderson 2007), 'through the concrete and highly visible form of agriculture' (Gascoigne 2002, p. 11). Conversely, those not amenable to or participating in improvement, such as the Australian savage, were assumed to be dying out. 'The extinct were thus the ones who had failed to realise their agency over nature' (Anderson 2007, p. 164).

Emerging Anthropocene narratives show strong connections to this modernist history. They have human actors who are dominant and powerful over, while being separate from, nature. As discussed in Chapter 1, Anthropocene origins are located not only with a human ancestor, but also very deep in time. The linear view of history and prehistory is inadvertently embedded within the dominant modes of visual representation – timelines, stratigraphic diagrams and exponential growth curves. The risk is a teleological view of human history in which the outcome (whether negative or positive) is inevitable, a visual trajectory further reinforced by the many exponential curves that direct our gaze to the future.

Scientific projections of climate change futures tend to be presented in a determinist way that can foreclose action, running the 'risk of nurturing uncertainty, apathy and inaction' (Brace & Geoghegan 2011, p. 292). Pahl et al. locate the origins of determinist futures in the modernist changes of the seventeenth and eighteenth centuries, when 'the future became increasingly conceptualized for the first time in its totality as a collection of interconnected possibilities' (2014, p. 378). This was closely linked to the new science of probability, which in turn enabled the rise of planning. The future was transformed into 'a calculable object' and 'became a new field of expert knowledge' (p. 378). As with the logics of progress discussed above, this is an understanding of time constituted by the experience of modernity itself: 'The extension and broadening of human expectations of control over the future differentiates the future constructed within modernity from that of previous epochs, in which posterity was seen as belonging solely to supernatural powers' (Pahl et al. 2014, p. 378).

The way in which the future is materialised in the present is addressed in some detail by Anderson (2010). Noting that there is much less geographic work on the future than on the past, he argues that a dominant problematisation is of the future as a disruptive surprise:[4] 'The risk is that we repeat a series of assumptions about linear temporality; specifically, that the future is a blank separate from the present or that the future is a telos towards which the present is heading' (Anderson 2010, p. 778).

Contesting teleological ideas of destiny, and linear ideas of progress, is especially important because of the extent to which they pervade a broader set of discourses, for example the many national narratives that drive economic growth (Jackson 2009), including narratives of national destiny.

On the other hand, the Anthropocene also contains the possibility of more contingent futures. It helps us understand that the conditions we associate with a 'stable' present are not at all stable when viewed at different timescales. A productive alternative view is De Landa's non-linear history, in which humanity 'liquifies' and 'solidifies' in different forms:

> if the different "stages" of human history were indeed brought about by phase transitions, then they are not "stages" at all – that is, progressive developmental steps, each better than the previous one, and indeed leaving the previous one behind. On the contrary, much as water's solid, liquid, and gas phases may coexist, so each new human phase simply added itself to the other ones, coexisting and interacting with them without leaving them in the past ... at each bifurcation alternative stable states were possible, and once actualized, they coexisted and interacted with one another.
> (1997, pp. 15–16)

My argument is not that we should forget about the past. Human life necessarily weaves many different temporalities together; our very bodies are the material legacy of billions of years of evolutionary change. It is rather to consider what it means to be open to less linear ways of thinking and practising time. As Brace and Geoghegan (2011) and Pahl et al. (2104) discuss, there are many challenges here for conceptualising and acting out more open futures, not the least of which is grounding them spatially in places to which people can feel connected. In the second half of the book we will explore some examples in greater depth. Here it is relevant to focus on two; the challenge of adapting a fixed national reserve system for changing conditions, and the everyday temporalities of climate change.

Reserve systems for the Anthropocene

In one sense debates about determined vs open futures have been around for a long time. Vegetation scientists have debated Clementsian vs Gleasonian ideas of plant succession for nearly a century, the former seeing change in directional terms towards a termination point, and the latter focusing on contingent combinations of individuals rather than a total community vegetation package (Pickett et al. 2009). There are both continuities and new challenges here, as the pace of ecological change accelerates. Biogeographical and ecological theories that assume equilibrium are inadequate for making predictions under novel conditions (Barnosky et al. 2012). Novel species assemblages not seen in combination before provide both conceptual and management challenges (Young 2014).

How well do framings based on past baselines prepare us for a future that will likely look very different? Directional trends in global climatic conditions suggest that many reserve systems are not sufficiently flexible to deal with climate change (Diffenbaugh & Field 2013), and it is widely argued that attention should focus on Conserving Nature's Stage, rather than the actors (Beier et al. 2015). This question is starting to be addressed in relation to the Australian National

Reserve System (including National Parks as well as other tenures whose primary role is biodiversity conservation). There is increasing recognition that climate change will challenge a national reserve system already stretched in various ways. It has had many additions in the last several decades, but is patchy and often made up of remnant land not useful for other purposes. Projections of future habitats, comparing future environments with the most similar current environment from anywhere on the continent, show widespread examples of the lack of current analogues (Dunlop et al. 2012). In other words, there is nowhere for some communities to move to. Lavau (2015) distinguishes between 'fortress' and 'fluid' conservation strategies, both currently co-existing in Australia. Fortress conservation, on which the National Reserve System is based, seeks to protect or restore historical conditions and communities of species. Fluid conservation, which recognises the challenges of a warming climate and the fragmenting landscape of different uses, seeks to improve permeability and hospitality. Lavau shows how wildlife corridors can be deployed to meet both ends; to allow species to move between safe fortresses, and/or to allow them to move in response to changing climatic conditions.

Prober and Dunlop get a bit more specific about what this means:

> it will be important to shift our focus from preserving species and current species composition, towards maintaining ecological and evolutionary processes . . . We may need to accept that gradually the giant tingle trees will disappear from the south-west forests and that snow gums may invade areas we currently value as alpine herbfield. But we can reasonably hope that current biota will be replaced by functioning, diverse ecosystems that effectively capture limiting ecological resources and provide ecosystem services, including beautiful places for human enjoyment.
>
> (2011, p. 2)

They do not use the word grief, but it is implicit in the injunction to accept the loss of the giant tingle trees (*Eucalyptus jacksonii*). Prober and Dunlop argue that we need a new philosophical discussion about what we value in biodiversity; they provide examples of possible goals, away from current species and combinations. We could aspire to maintain the evolutionary character of the Australian biota, with a target of say 85 per cent of plant species being endemic to the continent. We could move to a proximity principle in the reassembled communities, favouring proximate rather than distant species.

Lavau describes a situation in which both the politics of nostalgia and the politics of what she calls legacy currently co-exist. 'If the time horizon that most attracts the longing gaze of nostalgia is the past, for legacy it is the future. Furthermore, its gaze is more exploratory, its heritage practice more experimental' (Lavau 2015, p. 126). This relationship of co-existence is somewhat awkward; she notes it is likely to become more fraught as climate change creates additional pressures on the politics of nostalgia. The awkwardness can be seen in the Dunlop et al. (2012) report. It manages to sound fairly hopeful, arguing that the National

Reserve System is still our best hope of preserving species and ecological health, albeit it will need to become more flexible. This conclusion sits somewhat uneasily with its evidence about the lack of analogue conditions. In practice, the acknowledged limitations of the National Reserve System have already prompted change. The recognition that much under-represented biodiversity was on Aboriginal land led to the development of Indigenous Protected Areas in the late 1990s, and more recently partnerships between environmental NGOs (non-government organizations) and Indigenous groups have emerged (Moorcroft & Adams 2014). Future models are likely to need combinations of land tenures and engagement from many different community participants.

Temporalities of climate change

The foregoing discussion highlights the extraordinarily complex ways in which social life intersects with time. An emerging body of work examining the temporalities of climate change draws on a long heritage of thinking about social and ecological temporalities and how they interact. Reviews of this work discuss mismatches between the human mind, surrounding social dynamics and climate change (Brace & Geoghegan 2011; Pahl et al. 2014). Most commonly, people draw attention to the gap between our short-term behaviours and the long run of climate change: 'Thus humans are geared to prioritize short-term consequences of behavior and immediate futures. Evolutionarily, we are also used to close cause-effect relationships rather than those characterized by time lag and lack of contiguity' (Pahl et al. 2014, p. 376).

But Pahl et al. also identify an important conundrum in the opposite direction:

> people generally view climate change as long-term and slow, while important features of social life, such as economic relationships and our use of technologies, are increasingly structured by intensification and thus geared to immediate signals and short-term actions; in fact, it is actually very late in terms of action to slow or stop irreversible and accelerating climate change, and it is society that is moving very slowly.
>
> (2014, p. 382)

Either way, the mismatches between different temporalities are an important focus of research and consideration. As Bastian says, 'So while the clock can tell me whether I am late for work, it cannot tell me whether it is too late to mitigate runaway climate change' (2012, p. 25). She suggests we need to rethink the clock, away from a thing that measures time, to '*a device that signals change in order for its users to maintain an awareness of, and thus be able to coordinate themselves with, what is significant to them*' (Bastian 2012, p. 31, emphasis in original). There are many creative possibilities here to make different types of clocks, such as the One Hundred Months clock (which started in 2008 and signals the time available to take action against two degrees of global warming) and the clock of the Long Now (that would tick once a year, thus embodying deep time)

(Bastian 2012). More broadly, Bastian's work stimulates us to think of ways to express and embody the connections between different timeframes of climate change, and between humans and many non-humans.

Further, as geographers such as Massey (1999) have long argued, and as the example above of reserve systems also shows, times and places are deeply entwined and impossible to conceptualise separately. In their discussion of 'time stories' that anticipate sea level rise, Fincher et al. (2014, p. 202) identify and usefully review four ways that 'times in places have been conceptualised by geographers and others'.

- First, 'time is produced and understood materially, in the contexts that institutions and social mores regulate and that people inhabit' (p. 202).
- Second, there is a focus on everyday life, a scale often overlooked in conceptualisations of larger time-scales.
- Third, 'the time periods of past, present and future – those imagined, remembered and experienced times – form a central organising mechanism for thoughts and actions' (p. 203).
- Fourth, geographers have conceptualised time through 'explanations of how the ideas of the future influence contemporary meanings and practices' (p. 203).

Bringing these four together, Fincher et al. suggest that 'the everyday is the temporal site at which events and meanings at different temporal scales coalesce for people making sense of their situations' (2014, p. 203). The role of everyday temporalities was also identified by Norgaard as important in Bygdaby, where daily life 'is marked by a pronounced sense of the past' (2011, p. 112), connecting to the maintenance of cultural traditions, strong seasonal rhythms and the skiing heritage. She draws a link between the dissonant temporalities and the 'everyday denial' discussed in Chapter 2:

> Although there are many reasons why thinking about the consequences of climate change is not part of daily life in Bygdaby, one of them is clearly the disjuncture between the sense of time as it is normally experienced and the sense of time necessary to observe climate change or make its consequences seem "real".
>
> (2011, p. 114)

But these everyday temporalities can also provide sites of creativity in responding to the challenges of climate change. Fincher et al.'s (2014) study shows that the discrepancy is not always because the policy-makers and scientists have the correct temporalities and the community are resistant. In a study of the time narratives of mostly older residents in low-lying coastal communities in eastern Victoria – characterised in macro-analyses as a vulnerable population due to age and socioeconomic status – Fincher et al. show that people talk about *generational continuity*, in which value is attributed to family continuity, familiarity and the predictability of families being attached to the place over long periods of time. They also have a narrative of older age as being about *enduring and accepting* things that go with their choice to remain in this community as they age. On one hand you might expect such narratives to lead to resistance to change necessitated by actual or potential sea level rises and associated flooding. Quite the contrary – people talked about *the temporariness of inundation*, downplaying its significance as interrupting patterns of everyday life. (However, in demonstrating considerable resilience, they were not necessarily well placed to cope with very significant change.) They also complained about the *impermanence of the infrastructure*, and policy-makers' refusal to invest in infrastructure such as drains and pathways. In some ways the residents had a more realistic assessment of timeframes, connected to their own lifetimes. The local government is required by legislation to make decisions now for the projected higher sea levels of 2100, and is also clearly worried about investing in potentially stranded assets. Thus:

> The Victorian Coastal Strategy . . . [has] imposed upon local governments a requirement to implement actions in the present in order to avoid all risks into the future. If implemented as intended, this would bring forward the costs of adaptation over multiple generations onto present generations who, as we have explained, at present see little need and even less justice in this cost shifting exercise. Acting now to avoid all future risks is also inconsistent with most careful analysis of the temporalities in adapting to SLR, which call for dynamic, iterative, flexible, open-ended and sequential responses that evolve over time in light of new information and changing social conditions.
> (Fincher et al. 2014, p. 207)

Consequently, in the near term, residents 'consider themselves more vulnerable to the consequences of abrupt policy shifts in response to climate change than they do to climate change per se—a notion of vulnerability that stands in stark contrast to the expert framings of vulnerability' (Fincher et al. 2014, p. 208).

As we will discuss in Chapter 9, rethinking these everyday temporalities is one of the ways we can rupture linear time and also develop new, more sustainable, sociotemporal patterns. It also opens opportunities to revisit the past, and maintain or retrieve old practices.

Conclusion

The Anthropocene exposes the limitations of linear time. We cannot return to the past. But nor is the future any longer a location of hope, a time/space of improvement and progress, and of better lives for our children and grandchildren. Instead it has become an unimaginable time/space of potential catastrophe, of the undoing of modernity. While I will consider in the conclusion to the book the way in which the Anthropocene takes us into new kinds of time, the focus here is, rather, that it forces us to confront the limitations of how we use and understand time today. I have used the example of the Edenic Sciences, arguing they are prone to a particular kind of grief because of their adherence to past baselines. To reiterate from Chapter 2, the work of mourning is important and necessary. To some extent this work has gone unarticulated, or has been expressed in anxious, awkward ways, because the emotions of grief are difficult to reconcile within the sphere of dispassionate science. Bearing or holding grief provides a way for researchers to voice these feelings, while still doing good science. But we can also deal with loss more effectively if we are clear-eyed about those things that we never had.

These uncertain futures demand different and more mobile kinds of thinking about both space and time. Note here that I am not arguing we do away with the past in our imagination of the future, rather that we need to weave different temporalities together in new ways. The biodiversity conservation discussion suggests that one of the Anthropocenean characteristics deserving of greater consideration is openness to shifting and less linear temporalities. A less linear, more dynamic temporal framing may include rendering a more sophisticated sense of lifetimes and allowing for both long and short lead times between decisions and consequences (Stafford Smith et al. 2011).

We are living in the Anthropocene as we work on it. We necessarily have to work all this out as we go along, only partially with hindsight. As scholars we are in and of this history, and need to attend to the processes of category and thought construction just as much as the historical evidence of concern. A more contingent understanding of the Anthropocene is not only more historically accurate, it also provides more realistic and less fatalistic pathways to the future. If we are assuming humans will be part of the future, how can we articulate and enact the necessary creative human interventions? And how can we think the character of the human more constructively than has been done in the past? This is the theme of Chapter 4.

Notes

1 The slide series is preserved in the National Library of Australia collection (http://nla.gov.au/nla.pic-vn3885846). Last accessed 23 July 2015.
2 The Lake Pedder Restoration Committee (www.lakepedder.org/) have held a number of events to work towards the draining of the larger lake, and restoration of the original Lake Pedder (https://lakepedder.wordpress.com/). Last accessed 23 July 2015.
3 The stages were denial, anger, bargaining, depression and acceptance. As discussed in Chapter 2, contemporary psychiatric understandings see the grieving process as less predictable than this.
4 Note that Anderson focuses more on the possibility of oppressive rather than creative political outcomes in this imagining of risk and non-linear futures. He uses the examples of the war on terror, various biosecurity threats and climate change to argue that there are many parts of contemporary life where the future is an explicit influence on present actions: 'life is imagined as unpredictable, dynamic and non-linear ... Change cannot be understood as the linear outcome of past conditions or present trends' (p. 781).

> On the one hand, life must be constantly secured in relation to the dangers that lurk within it and loom over it. Life is tensed on verge of a catastrophe that may emerge

in unexpected and unanticipated ways. On the other hand, the securing of life must not be antithetical to the positive development of a creative relation with uncertainty. Liberal life must be open to the unanticipated if freedoms of commerce and self-fashioning individuals are to be enabled. Uncertainty is both threat and promise: both that which must be secured against and that which must be enabled.

(Anderson 2010, p. 782)

This argument may be convincing for the war on terror or bird flu, but I am not convinced in the implications for climate change – the arguments have not become that sophisticated yet. But Anderson is particularly concerned to see what kinds of relationships liberal democracies will preserve. 'To protect, save and care for certain forms of life is to potentially abandon, dispossess and destroy others' (Anderson 2010, p. 791).

References

Anderson, B. 2010. "Preemption, precaution, preparedness: Anticipatory action and future geographies." *Progress in Human Geography* 34(6): 777–798.

Anderson, K. 2007. *Race and the Crisis of Humanism.* London: Routledge.

Angus, M. 1975. *The World of Olegas Truchanas.* 6th ed. Hawthorn, VIC: Australian Conservation Foundation.

Aronson, J., C. Murcia, G. H. Kattan, D. Moreno-Mateos, K. Dixon and D. Simberloff. 2014. "The road to confusion is paved with novel ecosystem labels: A reply to Hobbs et al." *Trends in Ecology and Evolution* 29(2): 646–647.

Barnosky, A. D., E. A. Hadly, J. Bascompte, E. L. Berlow, J. H. Brown, M. Fortelius, W. M. Getz, J. Harte, A. Hastings, P. A. Marquet, N. D. Martinez, A. Mooers, P. Roopnarine, G. Vermeij, J. W. Williams, R. Gillespie, J. Kitzes, C. Marshall, N. Matzke, D. P. Mindell, E. Revilla and A. B. Smith. 2012. "Approaching a state shift in Earth's biosphere." *Nature* 486: 52–58.

Bastian, M. 2012. "Fatally confused: Telling the time in the midst of ecological crises." *Environmental Philosophy* 9(1): 23–48.

Beeton, R. J. S., K. I. Buckley, G. J. Jones, D. Morgan, R. E. Reichelt, D. Trewin. 2006. *Australia State of the Environment 2006.* Independent report to the Australian Government Minister for the Environment and Heritage. Canberra: Department of the Environment and Heritage.

Beier, P., M. L. Hunter and M. Anderson 2015. "Special section: Conserving nature's stage." *Conservation Biology* 29: 613–617.

Benson, D. and G. Eldershaw. 2007. "Backdrop to encounter: The 1770 landscape of Botany Bay, the plants collected by Banks and Solander and rehabilitation of natural vegetation at Kurnell." *Cunninghamia* 10(1): 113–137.

Brace, C. and H. Geoghegan. 2011. "Human geographies of climate change: Landscape, temporality, and lay knowledges." *Progress in Human Geography* 35(3): 284–302.

Cronon, W. 1996. "The trouble with wilderness; or, getting back to the wrong nature." In *Uncommon Ground: Rethinking the Human Place in Nature*, edited by W. Cronon, 69–90. New York: W. W. Norton and Co.

Crosby, A. W. 2004. *Ecological Imperialism: The Biological Expansion of Europe, 900–1900.* Cambridge: Cambridge University Press.

Davis, M. A., M. K. Chew, R. J. Hobbs, A. E. Lugo, J. J. Ewel, G. J. Vermeij, J. H. Brown, M. L. Rosenzweig, M. R. Gardener, S. P. Carroll, K. Thompson, S. T. A. Pickett, J. C. Stromberg, P. Del Tredici, K. N. Suding, J. G. Ehrenfeld, J. P. Grime, J. Mascaro and J. C. Briggs. 2011. "Don't judge species on their origins." *Nature* 474: 153–154.

De Landa, M. 1997. *A Thousand Years of Nonlinear History*. New York: Zone Books.
Diffenbaugh, N. S. and C. B. Field. 2013. "Changes in ecologically critical terrestrial climate conditions." *Science* 341: 486–492.
Dunlop M., D. W. Hilbert, S. Ferrier, A. House, A. Liedloff, S. A. Prober, A. Smyth, T. G. Martin, T. Harwood, K. J. Williams, C. Fletcher and H. Murphy. 2012. *The Implications of Climate Change for Biodiversity Conservation and the National Reserve System: Final Synthesis*. A report prepared for the Department of Sustainability, Environment, Water, Population and Communities, and the Department of Climate Change and Energy Efficiency. Canberra: CSIRO Climate Adaptation Flagship.
Fincher, R., J. Barnett, S. Graham and A. Hurlimann. 2014. "Time stories: Making sense of futures in anticipation of sea-level rise." *Geoforum* 56: 201–210.
Gascoigne, J. 2002. *The Enlightenment and the Origins of European Australia*. Cambridge: Cambridge University Press.
Green, D., J. Billy and A. Tapim. 2010. "Indigenous Australians' knowledge of weather and climate." *Climatic Change* 100(2): 337–354.
Green, D. and G. Raygorodetsky. 2010. "Indigenous knowledge of a changing climate." *Climatic Change* 100(2): 239–242.
Hinchliffe, S. 2008. "Reconstituting nature conservation: Towards a careful political ecology." *Geoforum* 39(1): 88–97.
Hobbs, R. J. 2013. "Grieving for the past and hoping for the future: Balancing polarizing perspectives in conservation and restoration." *Restoration Ecology* 21(2): 145–148.
Hobbs, R. J., E. S. Higgs and J. A. Harris. 2014. "Novel ecosystems: Concept or inconvenient reality? A response to Murcia et al." *Trends in Ecology and Evolution* 29(12): 645–646.
Hulme, M. 2009. *Why We Disagree About Climate Change*. Cambridge: Cambridge University Press.
Jackson, T. 2009. *Prosperity Without Growth: Economics for a Finite Planet*. London: Earthscan.
Lavau, S. 2011. "The nature/s of belonging: Performing an authentic Australian river." *Ethnos* 76(1): 41–64.
Lavau, S. 2015. "Climate change and the changing nature of conservation." In *The Future of Heritage as Climates Change: Loss, Adaptation and Creativity*, edited by D. C. Harvey and J. Perry, 111–129. Abingdon: Routledge.
Lefale, P. F. 2010. "Ua 'afa le Aso stormy weather today: Traditional ecological knowledge of weather and climate. The Samoa experience." *Climatic Change* 100(2): 317–335.
Lien, M. E. and A. Davison. 2010. "Roots, rupture and remembrance: The Tasmanian lives of the Monterey Pine." *Journal of Material Culture* 15(2): 233–253.
Lorimer, J. 2008. "Living roofs and brownfield wildlife: Towards a fluid biogeography of UK nature conservation." *Environment and Planning A* 40(9): 2042–2060.
Lorimer, J. and C. Driessen. 2013. "Bovine biopolitics and the promise of monsters in the rewilding of Heck cattle." *Geoforum* 48: 249–259.
Marris, E. 2014. "'New conservation' is an expansion of approaches, not an ethical orientation." *Animal Conservation* 17(6): 516–517.
Martin, R. J. and D. Trigger. 2015. "Negotiating belonging: Plants, people, and indigeneity in northern Australia." *Journal of the Royal Anthropological Institute* 21(2): 276–295.
Massey, D. 1999. "Space-time, 'science' and the relationship between physical geography and human geography." *Transactions of the Institute of British Geographers* 24(3): 261–276.

Miller, B., M. E. Soulé and J. Terborgh. 2014. "'New conservation' or surrender to development?" *Animal Conservation* 17(6): 509–515.

Moorcroft, H. and M. Adams. 2014. "Emerging geographies of conservation and Indigenous land in Australia." *Australian Geographer* 45(4): 485–504.

Murcia, C., J. Aronson, G. H. Kattan, D. Moreno-Mateos, K. Dixon and D. Simberloff. 2014. "A critique of the 'novel ecosystem' concept." *Trends in Ecology and Evolution* 29(10): 548–553.

Norgaard, K. M. 2011. *Living in Denial: Climate Change, Emotions, and Everyday Life*. Cambridge, MA: MIT Press.

Pahl S., S. Sheppard, C. Boomsma and C. Groves. 2014. "Perceptions of time in relation to climate change." *WIREs Climate Change* 5(3): 375–388.

Pickett, S. T. A., M. L. Cadenasso and S. J. Meiners. 2009. "Ever since Clements: From succession to vegetation dynamics and understanding to intervention." *Applied Vegetation Science* 12(1): 9–21.

Prober, S. M. and M. Dunlop. 2011. "Climate change: A cause for new biodiversity conservation objectives but let's not throw the baby out with the bathwater." *Ecological Management and Restoration* 12(1): 2–3.

Randall, R. 2009. "Loss and climate change: The cost of parallel narratives." *Ecopsychology* 1(3): 118–129.

Robbins, P. and S. A. Moore. 2013. "Ecological anxiety disorder: Diagnosing the politics of the Anthropocene." *Cultural Geographies* 20(1): 3–19.

Rose, D. 1997. "The Year Zero and the North Australian frontier." In *Tracking Knowledge in North Australian Landscapes: Studies in Indigenous and Settler Ecological Knowledge Systems,* edited by D. Rose and A. Clarke, 19–36. Canberra: North Australia Research Unit, ANU.

Simberloff, D. 2011. "Non-natives: 141 scientists object." *Nature* 475: 36.

Stafford Smith, M., L. Horrocks, A. Harvey and C. Hamilton. 2011. "Rethinking adaptation for a 4°C world." *Philosophical Transactions of the Royal Society A* 369: 196–216.

Trigger, D. S. 2008. "Indigeneity, ferality, and what 'belongs' in the Australian bush: Aboriginal responses to 'introduced' animals and plants in a settler-descendant society." *Journal of the Royal Anthropological Institute* 14(3): 628–646.

Weintrobe, S. 2013. "The difficult problem of anxiety in thinking about climate change." In *Engaging with Climate Change: Psychoanalytic and Interdisciplinary Perspectives*, edited by S. Weintrobe, 33–47. Sussex: Routledge.

Windle, P. 1992. "The ecology of grief." *BioScience* 42(5): 363–366.

Young, K. R. 2014. "Biogeography of the Anthropocene: Novel species assemblages." *Progress in Physical Geography* 38(5): 664–673.

4 More than human, more than nature

Perfect September dusk. Tide low, water still. We scrunch wet sand beneath our shoes, facing the rocky cliff opposite. A hooded human in a light-coloured coat enters left, across the water. The figure slowly walks, stretches, crawls, lies, curls – crossing our field of view, always just on top of the water (Figure 4.1). A black head at the waterline some metres in front is the only clue. The scene is accompanied by a flautist standing in the shallows, and later by readings from historical diaries that echo off the cliffs. We are all captivated, savouring the surreality along with the brackish whiff of the river and the thickening darkness. But we also wonder: how are they doing that?

Over dinner, some of the engineering and performative intricacies emerge. The raft was supported by milk crates with plastic bottles half full of water, calibrated to provide exactly the right amount of flotation. The tide and sunset times could

Figure 4.1 FLOAT, dancer Tess de Quincey, Siteworks 2010 at Bundanon, NSW. Photo: Heidrun Löhr.

be predicted, in order to bring the audience to the stage at the precise time. This art had a lot of science behind it. The stillness after a windy afternoon was perfect, but just lucky.

This was just one in a set of artistic performances at the Bundanon Siteworks Field Day.[1] The next day we were joined in conversations between artists and scientists about the future of food and water in a time of global warming. One thread in that conversation was distress and anxiety about climate change and the future, and the failure of society to take seriously what scientists are saying. For some scientists, artists were the messengers who would translate these arcane truths via stage, page or screen into something intelligible to the general community. Most of the artists were too polite to say so, but I don't think they saw their mission as being the public relations arm of science. There is plenty of empirical evidence to demonstrate that education about the facts is insufficient to generate changes in social practice. Notwithstanding differences of opinion about the purpose and role of the arts and sciences, Siteworks puts a bridge across the river dividing the two cultures. There is much to be learned from the conversation: how geomorphologists read landscape history in a stratigraphic section; how the performance artist Barbara Campbell camouflaged the audience and got them to think like a bird or a nest.

But framing an opposition between sciences and the humanities misses the point, wastes time and effort. This is not to deny the profound differences in how they go about things, nor the significant differences within each of what we call science and the arts. There is an eerie similarity in the contemporary incarnation of both science and arts: they frame the human as separate to the rest of nature. They are not facing each other across that divide; they are both facing the same direction, albeit equipped with different tools. At a time when human activities have become so deeply embedded in earth surface processes that even the molecular composition of the atmosphere bears our signature, the most urgent task for all fields of human endeavour is to reframe our relations to the more-than-human world. We don't need more modernist constructions like bridges; we need a different mode of relating to the river, a conversation about what it is like to all be standing on the same bank. In a sense we are all looking at the figure in the hooded coat (the performance artist Tess de Quincey). She is on the river, but not exactly in it. She is of the river, but not of the river.

The natural sciences took their subject matter to be the non-human world, excluding from their field of view various forms of human influence. They named their leading journal of science and technology *Nature*. Environmental thinking was framed around an ideal of human absence. The social sciences focused on human actors and agency, helping, in the words of the historian Tim Mitchell (2002, p. 52), 'to format a world resolved into this binary order' by constituting phenomena such as 'the economy'. Sociologist John Urry sees the emergence of his discipline as the product of the late-nineteenth-century historical moment when modernity – 'an emergent industrial movement-based capitalism' (Urry 2011, p. 7) – was transforming nature across Western Europe and North America. The humanities defined themselves around our species, taking scholarship into the human condition as their raison d'être. This binary formatting of the world developed a spatial logic – seen, for example, in the divide between nature and the city. When we preserve nature, we preserve it 'out there'.

Empirical evidence accumulating over the past few decades challenges each of these perspectives in different, albeit consistent, terms. Human difference was shown by evolutionary ecology and genetic research to be the contingent outcome of a few stray genes that seized the moment, and the behavioural consequences that reinforced various moments, rather than a divinely ordained status within creation. Archaeological and palaeoecological evidence demonstrated in Australia, as elsewhere, that 'nature' had been neither stable nor pristine. Climates became colder, hotter, wetter, drier. Sea levels rose and fell, creating new configurations of land and seascape. Humans were participants in Australian ecosystems for tens of thousands of years: burning, hunting, choosing some plants over others, transporting plants and animals to different parts of the landscape in the course of everyday life.

Critique of the human/nature binary has been a feature of scholarship in the humanities and social sciences over the past twenty years. There have been widespread attempts to unsettle and dismantle it, as well as necessary work in analysing its extraordinary resilience and embeddedness in our thinking and institutions.

So, the natural sciences are discovering people, and the human sciences are considering the non-human world more systematically. Scholars are still groping with the implications of these findings and new perspectives. This is the challenge for the twenty-first century, and it is not easy. We talk of the 'posthumanities' to signal the decentring of the human subject, and the 'more-than-human' to acknowledge both the pervasiveness of human influence and its interaction with non-humans (plants, animals, rocks, weather). Even the boundary of life is not as clear as it once was, so the previously neat categories of technology, or materials, are giving way to more lively configurations of matter. The clumsiness of these terms illustrates the difficulty we all have in shifting our modes of thought, language and practice.

Here then is the key problematic of the book. Central concepts whose empirically constitutive data demands a radical rethink of how we approach the world are themselves saddled with a separationist view of the human. The Anthropocene is understood to demand new thinking around – and responsibilities for – planetary stewardship, but many of its manifestations perpetuate a modernist understanding of human domination over nature, for example in ideas about planetary governance. This applies equally to climate change itself:

> The discursive separation of climate and society leads inexorably towards the representation of climate change as an exogenous force that manifests itself in the form of external shocks to an otherwise independent society. This conceptual framework, I believe, is both limited and limiting. It is perhaps ironic that the discourse of adaptation is founded on intrinsically dualistic foundations just as the notion of 'anthropogenic climate change' appears to challenge them.
>
> (Taylor 2015, p. 11)

This separation flows into climate adaptation, the seeming naturalness of which:

> stems from the way its framework replicates the essential dualism between society and nature that is deeply embedded in Western social thought. Framed within this Cartesian ontology, the idea of adaptation builds from the notion that climate and society are two distinct domains or systems that engage in a series of external influences.
>
> (Taylor 2015, pp. 189–190)

The concept of 'human impacts' emphasises the moment of collision between two separate things, and that of 'social and ecological systems' is also about stitching together two pre-constituted entities. The question of whether humans are natural is entwined with that of whether they can unbalance nature, and whether their actions are good (Lynas 2011). To argue that we need to stitch back together – or to build bridges across – systems or ways of thinking serves only to underscore this ontological separation, rather than to overcome it.

There are many contradictions here, and it is important that our discussions acknowledge our contradictory positions. But the main divisions are not between science and the arts. For those of us who grew up with Western thinking, our most profound and important challenge is the challenge of reconceptualising human relations to the more-than-human world. It will not occur as a purely cerebral activity, but as a process of engagement with the dilemmas of everyday practice. To undo the destructive practices of modernity, and reconstitute them into something better, we will need everything in the Enlightenment toolbox, science and arts included. But they will be most effective plunging into the river together, rather than attempting to bridge it.

In this chapter I aim to help seize the potential for the Anthropocene to contribute to that radical rethink, rather than resediment Cartesian modernity. The chapter focuses on the figure of the human. Although I do not have conclusive answers, an important start is to identify those tensions and contradictions. My perspective is posthumanist, in the sense defined by Sundberg, that it 'refuses to treat the human as 1) an ontological given, the privileged if not the only actor of consequence and, 2) disembodied and autonomous, separate from the world of nature and animality' (2014, p. 34). I advocate a relational and associationist perspective on the human, entirely consistent with my understanding of evolutionary history. I want to reclaim a sense of human which has always been a variable container; we can have a contingent and also variable understanding of human-ness. But this perspective also has a key weakness; it does not necessarily acknowledge and deal well with human power. And I do want to maintain a sense of human power and responsibility to contribute healing interventions to the earth system. The point is then, not that we can escape the contradictions, but to help us live in multiplicity.

Such a rethink would have several characteristics, or aspects that must be considered. These include multiple configurations of the human and its difference from and between; differentiation of non-humans (I use the example of plants);

causation and intervention by association and relation; and considering human pervasiveness and power in earth systems.

Australian evidence and perspectives provide an important place from which to contribute to these radical rethinks. I demonstrate this by first revisiting Kay Anderson's work, and then going on to the question of relational, more-than-human and Bawaka Country politics. Of course an Australian perspective does not solve all the issues, but it does provide some unique insights into the ways our understandings of the human are constructed.

Multiple configurations of the human

An important step is to recognise that there are many configurations of the human in contemporary debates; we have 'to think of human agency over multiple and incommensurable scales at once' (Chakrabarty 2012, p. 1). Chakrabarty identified three configurations:

> the universalist-Enlightenment view of the human as potentially the same everywhere, the subject with capacity to bear and exercise rights; the postcolonial-postmodern view of the human as the same but endowed everywhere with what some scholars call "anthropological difference"—differences of class, sexuality, gender, history, and so on. This second view is what the literature on globalization underlines. And then comes the figure of the human in the age of the Anthropocene, the era when humans act as a geological force on the planet, changing its climate for millennia to come.
>
> (2012, pp. 1–2)

Chakrabarty's disjunction (attention to anthropological difference) is placed against a global warming in which humans are 'constitutively one – a species, a collectivity whose commitment to fossil-fuel based, energy-consuming civilization is now a threat to that civilization itself' (p. 2). For Chakrabarty, these figurations of the human just have to sit together: 'One cannot put them along a continuum of progress. No one view is rendered invalid by the presence of others. They are simply dis-junctive' (Chakrabarty 2012, p. 2).

These multiple figurations of the human are also pervading archaeological time. The unitary model of human evolution, in which *Homo sapiens* is the last of many ancestors standing, is being challenged by new archaeological and DNA discoveries. These include the findings that Neanderthals contributed 1–4 per cent of contemporary European and Asian nuclear DNA, and that a genetically different population, dubbed the Denisovans, was living in Siberia at times contemporaneous with modern humans (Gibbons 2011, Hawks 2013, Yusoff 2013). 'This means *H. sapiens* mixed it up with at least two different archaic peoples, in at least two distinct times and places' (Gibbons 2011, p. 393). The mooted contemporaneity of *H. sapiens* and *H. floresiensis* on Flores (Brumm et al. 2006, Roberts et al. 2009) provides another example. My focus here is not what this means for the specifics of the human migration story, fascinating as that is.

Given the rapid pace of research, the specifics of any part of this story are likely to change considerably in the next few decades. Rather it is on the way the unitary (and progressivist) model of human evolution is being broken down. Yusoff (2013, p. 787) goes so far as to argue that 'these recent discoveries have reconceptualised humanity as interspecies'. Certainly the concept of species itself is a contested one (Gibbons 2011), as is the concept of 'modern human' (Stringer 2012). As Yusoff argues, we increasingly have a picture of 'hominin evolution as temporally, sexually, and geographically differentiated in their migration and forms of territorialisation' (Yusoff 2013, p. 787). Even just the mental picture that humans were not alone in their world provides a strong challenge to our deeply embedded understandings of the human self and its separateness. On the other hand, that species aloneness is itself confronting; 'it can be emotionally difficult to absorb the radical contingency of humanity' (Maudlin 2013).

As Castree (2015) has shown, the human/nature distinction is maintained in many arenas of contemporary life. It is not easy to disembed. Nor do we necessarily want to get rid of any concept of human difference, but can we understand it in a non-hierarchical way? Plumwood's (2006) theory of mutuality, which acknowledges both continuity and (non-hierarchical) difference between humans and non-humans, is helpful here. However, in contrast to posthumanist approaches, Plumwood retained the concepts of nature and natural systems. She rejected concepts such as 'naturecultures' that aim to implode the distinction between the two, arguing that rejecting hyperseparation is not to reject difference or distinguishability:

> What is lost when we refuse to acknowledge difference between nature and culture, or when we accept an idealist or social constructionist reduction of nature to culture? There may be a range of situations in which they are hard to separate, but there are an important range of others in which recognizing their difference is crucial.
>
> (2006, p. 144)

Such a view has resonance in Australia, where the battle to include humans in conceptualisations of nature is not yet won. Many Indigenous communities are still struggling against environmental management regimes that ignore or erase their presence in the landscape. In such a context they can find the concepts – both anthropocentric and dualistic – of 'cultural landscape' or 'cultural flows' (Weir 2009) to be powerful political tools. This unfinished business also affects settler Australians who still broadly understand themselves as outside nature (Head & Muir 2007). We can thus accurately conceptualise many Australian contexts as needing a more-than-nature approach, rather than a more-than-human one.[2]

We return below to the ways we can learn from Indigenous ways of seeing, an important reminder of the profound differences between different groups of humans in their environmental relations. So, while the Anthropocene may be a crisis facing humanity as a whole, it will affect us all differently. As many have

Dehomogenise the non-human: the example of plants

Over several years now, I have been exploring the nature of plants and plantiness with colleagues Jenny Atchison and Catherine Phillips (Head et al. 2015). We noted that although the agency of plants is increasingly demonstrated in empirical research, scholars were yet to fully respond, for plants, to Lulka's (2009) call to attend more carefully to the details of non-human difference. He argued that there is a residual humanism in the use of the hybridity concept when 'nonhumans'[3] are lumped as a singular entity, and that it is important to pay attention to the differences in non-human difference.

In this analysis, plants *emerge* as an assemblage of shared differences from other beings, where common capacities manifest in different material form. We understand plant capacities as relational achievements, the relations in question enacted with the sun, water and soil, and also often with humans and other animals. These are contingent relationships in which categories and forms should not be reified, even though they may be stabilised for very long periods of time. Given the diverse ways in which the concept of relationality is understood in the literature, we emphasise three ways in which we see such capacities as relational. First, we do not assume humans have to be part of the relations, although they often are. Second, relationality can be intra-organism, referring to the processes constituting that organism. It is not only about external relations. Third, relationality is not a synonym for dynamism and fluidity. While it has those qualities, it can also congeal and solidify (in De Landa's 1997 terms, *intensify*) in forms and processes that persist over long periods of time. In the case of plants, these are extremely long periods of evolutionary time.

The four distinctive capacities of plants include: 1) the particular materialities deriving from the performance of photosynthesis and associated material consequences; 2) moving independently of humans; 3) sensing and communicating; and 4) having flexible bodies. Our argument then is not only that plants have agency (a point well made by others) and subjectivity (a more controversial proposition), but also that focusing on plant worlds shows how human-centred our conceptualisations of agency and subjectivity have been.

Plants help to rethink *bodily* difference beyond the human (and the animal), extending explicitly feminist theories that have made significant ground in the recognition of non-human difference (Hird & Roberts 2011). Plant bodies challenge our understanding of individual and collective bodies. One example is the plant capacity for sporic meiosis, where two morphologically distinct bodies alternate in the life history (Graham et al. 2000). This is further complicated by growth and development; a process in plants of not simply enlarging juveniles (as animals do) but building assemblies or 'confederations' of new and old members (Firn 2004). The variability of plant bodies helps us think more laterally about the relationship between the body and the self. Part of the value of such an approach is that it helps us reflect on the qualities of human-ness. As Kohn argued, the reason this work matters is not just that it gives 'voice, agency or subjectivity to the nonhuman—to recognize them as others, visible in their difference—but [that it forces] us to radically rethink these categories of our analysis as they pertain to all beings' (Kohn 2010, quoted in Kirksey and Helmreich 2010, pp. 562–563).

Plants challenge thinking about agency and subjectivity against a human norm; in contrast to many animals, plants are so different from us that we are not at risk of confusion. The point is not that plants possess agency, but that they enact distinctive agencies – sun eating, mobile, communicative and flexibly collective.

We return to some specific examples of associative relations – including adversarial ones – between plants and humans in later chapters. While we are thinking about difference, it is important not to lose sight of the role of non-human others in the actual bodily associations that constitute the human. Plants underpin our food supply and contribute to the air we breathe. As Yusoff (2013) reminds us, our constitution as human is not only biotic but also mineralogical, carrying ancestral traces from deep in earth history.

From impacts to relational interventions

Relational concepts of agency challenge linear understandings of causation. This both opens up and complicates the role of human intervention in earth processes, whether we are trying to understand the past or act in the future. The emphasis on the moment of collision between two separate entities (the 'impact' of 'humans' on 'climate') has favoured historical explanations that depend on correlation in time and space, to the detriment of the search for mechanisms of connection, rather than simple correlation. This is particularly important to how we think about the future, since removal of the 'human' is presumably not our solution of first resort. Nor, as Chakrabarty argued, is there any overall 'human' that can act together:

> The fact that the crisis of climate change will be routed through all our "anthropological differences" can only mean that, however anthropogenic the current global warming may be in its origins, there is no corresponding "humanity" that in its oneness can act as a political agent. A place thus remains for struggles around questions on intrahuman justice regarding the uneven impacts of climate change.
>
> (2012, p. 14)

Instead it is necessary to articulate diverse locations of generative political action that can create traction in the same direction. Responses to climate change will require human action, in association and assemblage with many other types of agency. This complicates the field of view, but also helps us imagine diverse points of intervention.

These will include restorative human influences (Ellis 2013). In some situations, anthropogenic loss of native species and anthropogenic introductions interact to increase species richness in regional landscapes, even while global biodiversity is thinning (Ellis et al. 2012). Within the growing field of urban ecology there is increasing recognition that the co-option of human actors is likely to be crucial to biodiversity conservation. Rudd et al. (2002) have shown that the optimal configuration of habitat networks in Greater Vancouver would need to include backyards. Biodiversity conservation in both the fragmented landscapes and the large protected areas of the Sydney Basin will be at risk unless humans can be re-imagined and co-opted as active co-constructors of this nature rather than solely as threats to it (Head & Muir 2006). Now that, for the first time in human history, most of the world's population lives in cities, discussing these issues is of wide international relevance.

The Australian contribution

To more fully illustrate my argument about the contingent constructions of humanness, I return to the specifics of the Australian role in these debates. My focus is on two particular manifestations. In Anderson's work, her rereading of the nineteenth century provides important lessons for the twenty-first century configuration of the human. In the encounter between academic geography and Indigenous ontologies we have to face the consequences of radical difference.

As mentioned in Chapter 1, the Australian colonial encounter collided with, and helped reconfigure, the emerging understanding of the human: 'The humanist edifice of "the human" in western European thought, was thrown into crisis in the early-mid 1800s in confrontation with radically divergent modes of life in far-flung places' (Anderson 2007, p. 24).

Anderson argues that 'evolutionary theorising was not so much applied to Australia, as *constituted* out of materials from a time/space location where the archaic human first began lifting himself up' (2007, p. 176). In other words, '[t]he colonisation of that continent [Australia] from the 1780s onwards did not proceed unfalteringly with racialised stereotypes of its indigenous people *in hand*' (Anderson 2007, p. 27, emphasis added). Specifically, the encounter with Indigenous Australians exposed 'the fragility of a power-base that staked its logic in the idea of human separateness from nature' (Anderson 2007, p. 27):

> Note here that humanness is presumed to consist in a movement not only beyond external nature, but crucially, also, human animality. In humanist thought – and especially since its Christian strand became intricated with modernity during western Europe's Enlightenment – humanness has been presumed to entail a transcendence of nature conceived as: first, a non-human sphere of animals and environment that is 'external' to people; and/or second, a corporeal nature that is thought of as animal-like which resides within human beings.
>
> Regarding the first such notion, of a nature that is considered 'external', one can note the significance attached to nature's cultivation in the western cultural tradition's story of the human transformation of the earth ... With

tools, crafts, arts and machines, people made good the deficiencies of nature. Nature was to be *cultivated*. And, as the etymology of this word implies, it was in the labour of cultivation that the cultural rather than natural character of the human was articulated.

(Anderson 2007, p. 8, emphasis in original)

The role of agriculture, and its apparent absence in thinking about the human, is discussed further in Chapter 6. Regarding the association with, and separation from, animality, Anderson is keen to examine the process by which such understanding was materially constructed. In so doing she is critical of 'critical race studies':

what has been missed in such standard liberal critiques of these familiar and persistently pernicious tropes, is that the figure of 'the human' – into whose fold it is judged that indigenous people should have been admitted – cannot itself be assumed as ontologically given, neither in the past nor present day.

(2007, p. 11)

In two more recent pieces (Anderson 2014, Anderson & Perrin 2015), Anderson makes connections between this deconstructive examination and the persistent human exceptionalism of the present day. Anderson and Perrin (2015) ask, what is the material basis of contemporary human exceptionalism? (They do not mention the Anthropocene but they could have.) They do this against the background of the 'increasingly 'materialist' concern within the humanities (and social sciences) to move beyond the legacy of a narrow, humanist conception of culture as something separate from, and elevated above, the natural world' (2015, p. 1). They take issue with 'the contention that those qualities commonly regarded as differentiating and elevating humans from nonhumans are mere fantasies' (p. 3), instead attempting to trace the material construction of such ideas:

It is, therefore, in extending the insight that 'everything is material' to humanism itself that our aim here is to 'open up' the (radically ventured yet profoundly under-examined) thesis of human exceptionality, in order to see precisely how it has been assembled.

(2015, p. 4)

At the beginning of the nineteenth century 'the comparative anatomists of the period elaborated an explicitly scientific—and, indeed, materialist—account of human exceptionality' (Anderson & Perrin 2015, p. 6). In other words, humanism was always a materialist configuration, even as it claimed to focus on putatively non-material things such as the soul. Following Latour, their materialist understanding 'precludes dismissing certain things—like the soul or the mind— as products of mere thought, language, belief or fantasy' (Anderson & Perrin 2015, p. 8).

Anderson and Perrin connect their historical analysis to contemporary debate via Clive Hamilton's (2010) critique of geo-engineering and human hubris:

> We accept that on the point of humanity's environmental impact Hamilton's concern is well placed. But to consider this 'objectification' [of nature] as a kind of unwavering ideology, as an unswerving commitment to some invariant belief, is to grant it, not just an implausible constancy, but a dangerous consistency. It is, then, to obscure the very historicity of humanism. We urgently need to overcome the still pervasive idea that the value of human reason, meaning, knowledge-making and creativity lies in rising above our worldly—and indeed our animal—existence.
>
> ... this figure of the nature-transcendent human must, we have argued here, be understood not as an otherworldly fantasy, but as a worldly, and for this reason an always-contingent, always-unstable, production. A materialist engagement with—rather than disengagement from—the idea of human exceptionality is, therefore, vital for a humanities tuned to a planet under pressure.
>
> (Anderson & Perrin 2015, pp. 11–12)

Anderson and Perrin remind us, like Gibson-Graham (2008), that because our understandings of the world are materially constituted in situated ways, they can also be reconstituted differently. So we can create a more open and contingent human for the Anthropocene, one (or many) who can act restoratively and collaboratively rather than destructively.

In my second example, one of the models for an alternative constitution of human relations to the more-than-human world is provided by Indigenous ontologies. In this example I analyse some of my own reaction to Bawaka et al.'s recent series of papers in which they foreground the agency of Country. First I provide some history about Australian geography's recent engagement with the Indigenous presence. I would argue this has been pervasive, even for those who have not explicitly worked on Indigenous issues.

We start with important work by Richie Howitt, in which he drew Australian empirical material into conversation with discipline-wide concepts of scale, borders and the other (1993, 2001, 2002). Howitt drew on an emergent body of scholarship in which Australian geographers, anthropologists and others had been encountering Aboriginal communities, views of the world and engagements with land, water and fire (see also Howitt & Jackson 1998). The widely quoted Australian Indigenous conceptualisation of Country, such as in Rose's (1996) *Nourishing Terrains*, understands Country as sentient. Rose writes of Australian Indigenous philosophies:

> Subjectivity, in the form of consciousness, agency, morality and law, is part of all forms and sites of life: of non-human species of plants and animals, of powerful beings such as Rainbow Snakes, and of creation sites, including trees, hills and waterholes. Nourishing terrains are sentient.
>
> (1999, p. 178)

In contrast to a perspective informed by Country, Howitt argued that Australian landscapes as managed by settler governments 'are plagued by multiple boundaries that seek to divide and subdivide places, people and resources into manageable units' (Howitt 2001, p. 233). He took issue with the tight boundedness of frontiers and borders, through which colonial approaches to the landscape continued to be reinscribed in the present, offering challenges to political perspectives of both Left and Right. Howitt suggested instead the constructive and creative use of edges (in an ecotone sense), giving us metaphors for the co-existence of Aboriginal and Euro-Australian engagements with land and resources. He encouraged the geography discipline to go beyond its role in creating geographies of exclusion, and move towards openness, multiplicity and coexistence.

With Sandie Suchet-Pearson (Howitt & Suchet-Pearson 2006), the argument was extended into a critique of Western discourses of community and environmental 'management' that presumed the possibility of human intervention and control. In this and related work (Suchet 2002) they started to imagine what 'ontological pluralism' might look like, and how it might take shape through a focus on process in environmental and development engagements. Further, they argued there is also a temporal frame of reference in the Eurocentric dominance of management discourses, which 'orientate thinking towards a linear narrative – with a unidirectional, progressive, controlled movement towards a coherent strategic target presumed desirable' (Howitt & Suchet-Pearson 2006, p. 332).

In these respects, the Indigenous engagement helped to constitute the Australian expression of posthumanism. Note here that my perspective is quite different to that of Sundberg's (2014), and to a lesser extent Jackson's (2014), postcolonial critique of more-than-human thinking. They argue that more-than-human approaches critique the nature/culture binary in universalist terms, rather than acknowledging the Western centredness of the binary. These approaches, argues Sundberg, should go much further in acknowledging or engaging with already existing non-dualistic worldviews, such as those of Indigenous societies (Sundberg 2014). In a critique of her own work, as well as that of leading posthumanist scholars, Sundberg argues, 'the "more-than-human" methodologies currently elaborated in posthumanist geographies are but one approach in a world of inclusive or non-dualist frameworks, such as those articulated by Indigenous scholars' (2014, p. 36). I think she misreads the history of posthumanist thought, at least in relation to Australia. In the Australian experience what came to be called more-than-human perspectives in geography emerged, among other influences (including the Anglo-American centre of the academy!), from the necessary confrontation and engagement with Indigenous politics and ontologies in the 1980s and 1990s. This is too detailed an argument to have fully here (but see the references above, among many others). However imperfect and stubbornly colonial was that engagement, it was this confrontation more than any other that undid the dualisms of nature and culture. It is even arguable how influential it was in key northern hemisphere texts such as Whatmore's (2002) *Hybrid Geographies*, which has an Australian chapter. So Indigenous agency, indeed Country itself, has already been influential.

This is not to say that we have solved all the issues, or that we have decolonised the discipline completely. Some of the ongoing ontological dilemmas can be considered via the important work of Sandie Suchet-Pearson. She has worked for a number of years in collaboration with two other non-Indigenous academics (Sarah Wright and Kate Lloyd) and several generations of Indigenous women (including Laklak Burarrwanga, Ritjilili Ganambarr, Merrkiyawuy Ganambarr-Stubbs, Banbapuy Ganambarr and Djawundil Maymuru) elders for Bawaka Country in northeast Arnhem Land. They describe themselves like this: '*We* are an Indigenous and non-Indigenous, human and more-than-human research collective' (Bawaka Country et al. 2015a, p. 271, emphasis in original). Cue photo of sprawling people (adults, children and baby), a kangaroo that also sprawls, quite at home, sand, canvas, water, rocks, swings, football:

> Most importantly, *we* is Bawaka itself. Bawaka is an Indigenous homeland on the water of Port Bradshaw off the Gulf of Carpentaria in the north of Australia. The notion of *homeland* encompasses not just the land itself, but also all the beings (human and nonhuman, corporeal and non-corporeal), processes, affects, songs, dreams and relationships that co-constitute Country. Bawaka is our coming together, our co-becoming.
> (Bawaka Country et al. 2015a, p. 271, emphasis in original)

There are a number of publication outcomes from this work. Here I focus on those in which Bawaka Country is provocatively positioned as the lead author, as an ethical imperative (Bawaka Country et al. 2013, Bawaka Country et al. 2015a, 2015b). It is partly my reactions to this work that I want to discuss. In Bawaka Country et al. (2015b), embodied narratives of digging yams are woven through the discussion. The aim is explicitly not a definitive ethnography: 'Rather, certain insights into gurrutu [the Yolŋu kinship system] help us grapple with some of the challenges facing academics as they try to understand place/space in relational, material and more-than-human ways' (2015b, p. 3). In places the writing is quite lyrical in describing the shared Bawaka world of humans and others: 'This world is differentiated, often incomprehensible, but the air still passes between us, the meaning of the world is still co-defined, the sound of the gukguk [a type of pigeon] is woven into the sensory universe, the relationality remains' (2015b, p. 9).

The theoretical claims in this paper are both modest and all-encompassing:

> We have looked to Bawaka not as a case study of a 'different' way of understanding space, but for what it can teach us about how *all* views of space are situated, and for the insights it offers about living in a relational world. *Gurrutu* helps geographers and others to get at the bounded/unbounded, contingency/coherence tensions within geographers' debates about relationality by highlighting that relations are always co-becoming yet have diverse definitive underlying relations that guide and structure their co-becoming (e.g. Yirritja/Dhuwa, yothu/yindi). . . . It denies humans the fabrication that they may stand aside, act *on* each other or a discrete non-human environment.

It places us all squarely within an ethics of co-becoming and demands that we attend to the connections that bind and co-constitute us.

(2015b, p. 15, emphasis in original)

This research centres itself within Indigenous epistemologies and ontologies. They align themselves with Karen Martin's suggestion that 'Indigenist research occurs through centring Aboriginal Ways of Knowing, Ways of Being and Ways of Doing' (Martin 2003 in Bawaka Country et al. 2015a, p. 274). It also places itself firmly within a more-than-human mode of operation. For these authors, 'research and research methodology in this context is clearly a more-than-human practice' (Bawaka Country et al. 2015a, p. 274). This suggests the authors see a strong overlap, possibly even equivalence, between the two approaches.

The Bawaka papers are, among other things, an invitation to a conversation, and it is in this spirit that I respond to their excellent provocations. For me, a central question is what do we do with these 'ontological frictions' (Palmer 2015, p. 19)? Do they sit easily beside one another? If they were put on a Venn diagram, what would they look like? Both Bawaka Country et al. (2015a) and Weir (2009) see a strong fit between more-than-human (or what Weir calls 'amodern') approaches and Indigenous ontologies. They would probably have overlapping circles. In contrast, I read Jackson (2014) and Sundberg (2014) as arguing that Indigenous ontologies should be brought into a bigger, more-than-human umbrella. Jackson takes issue with Chakrabarty's (2012) notion of disjuncture, arguing that it reproduces, indeed is dependent on, 'an implicit eurocentric and colonialist division between universal materialist nature, and particular symbolic systems of culture and value' (2014, p. 74). Jackson goes on to invoke Indigenous ontologies: 'In particular, many aboriginal and indigenous ontologies may be invoked as far more relational means to understanding material immanence' (p. 82). It may not have been intended in either of these papers, but the implication that more-than-human approaches need to absorb all kinds of relational or non-dualist ontology can be construed as a kind of colonising or universalist move in itself.

I am entirely in agreement with several dimensions of these arguments. Western posthumanism, particularly its Anglo-American expressions, could do much more to situate itself in place and time rather than carrying on the conversation in a way that implies a universal 'we'. Western posthumanism could constructively draw on the many examples of relational or non-dualist ontologies from Indigenous and other societies of the Global South.

Western posthumanism could decolonise its practice, and its relationship with Indigenous communities, by attending more carefully to the kinds of ethical issues discussed in detail by Bawaka Country et al.

For me, more-than-human perspectives should provide neither an umbrella over everything, nor an easy overlap with Indigenous ontologies. Rather, I see more-than-human perspectives as distinctly Western ones, and we should acknowledge their situatedness as such. I need them to just sit beside each other, rather than one absorbing the other, or them completely overlapping. For as

Bawaka Country et al. point out, 'for non-Yolŋu, we would argue that it is simply not appropriate or possible to fully live the ontological realities of *gurrutu*' (2015b, p. 15). Sitting beside one another is still to be in relation; as I have argued above, they have been in relation in Australia for several decades now.

Further, it is hard to get beyond this discussion as a human-centred process: 'Co-becoming is *our conceptualization* of a Bawaka Yolŋu ontology within which everything exists in a state of emergence and relationality' (Bawaka Country et al. 2015b, p. 2, emphasis added). However such an ontology has a place for non-human others, it comes to outsiders through a human lens of understanding and explanation. (I accept that insiders would understand it as unmediated.) So it is a view from somewhere, as is the relational ontology I want to defend from within a Western scientific framing.

Perhaps my biggest question is 'what can we do about it'? If we respond with care and responsibility, how does it help us fix things? Bawaka Country et al. (2015b) acknowledge that power is a thorny issue in these perspectives. Chris Gibson and I (Head & Gibson 2012) pondered this question in relation to how far Weir's (2009) amodern thinking gets us along a practical path of healing for the River Murray. Amodern thinking may mean we have to accept things we cannot fix, and first grieve appropriately (as Weir suggests in part of her final chapter called 'Acknowledging ecocide', and as I discussed in Chapter 2). We could then move on to living with the new and changed reality, such as that of a dead river. I take the point that it is first important to just be, to listen and to attend deeply. And, following my argument in previous chapters, this grief must be borne and witnessed. But it must also be carried with us as we intervene; I am not yet ready to completely abandon the modernist desire to fix things.

A sort of 'we' opens up

To summarise: my argument in this chapter is not to bring things to a close, but to situate both ourselves as scholars and readers and the unstable human character in space and time. Relations can solidify into particular forms and processes and endure over evolutionary and shorter timescales. They can also be disrupted, fall apart and be reconfigured. To attempt a summary is also to articulate the many tensions that attend my path. As Instone and Taylor (2014) argued, it is important to stay with the mess of our inheritance. Indeed we cannot do otherwise. But in doing so, they suggest, a sort of 'we' opens up – risky attachments of possibility between humans, and between humans and others.

I have situated my own perspective in its Enlightenment heritage, albeit one that aims to 'become differently modern' (Head & Gibson 2012). My perspective on relationality comes from an understanding of evolutionary history and its many contingencies. I see the Anthropocene as providing a window in which the links between posthumanism and the non-linear natural sciences can be developed. As the introductory example of the Bundanon River art suggested, this is explicitly not about stitching together preconstituted entities. In exploring some aspects of how my Australian perspective has been constituted, I acknowledge its relationality with, among many other things, Indigenous ontologies. Inasmuch as Bawaka Country et al. offer an invitation and hospitality, it behoves us to think much more carefully about what we can 'learn' from the encounter, and what the different contributions of these two different ontologies (neither of them in fact monolithic) might be.

In the Australian example, non-Indigenous academics and others have learned much from Indigenous perspectives about deep affiliation and responsibility to land and country. We have become aware of the importance of 'situating self in relation to community affiliation and place' (Sundberg 2014, p. 36), and understand better how the present weaves together a number of temporalities. We have learned about humility in relation to Country; we perhaps have not yet learned humility itself. We have learned about attending deeply to companions and others:

> To attend deeply is not a frivolous exercise yet neither is it something that should be cast in the too-hard-basket, as something strictly for Indigenous people. The need to attend, to listen, to feel, to respect and to practice methodologies that challenge human superiority is an ethical question, one that touches all who research.
>
> (Bawaka Country et al. 2015a, p. 277)

However, we are wary of appropriating Indigenous culture and knowledge, and we wonder what it means for us to attend to Country. And where we might do this.

***

In the depths of winter the bush has withdrawn into itself. There is a new openness – sightlines not visible when the full flush of growth comes a few months later. But things are not still. The golden balls of the one we call wattle are bursting into bloom. Their trees can be annoying and the wood does not burn well, but I defy anyone to not be cheered by the sight of wattles in flower sprinkled through the forest. The one we call daffodil also pops up at this time, planted by someone else around the old house. Its stalks have been folded by this week's strong winds and I pick a few of these for the vase. The one we call banksia has been in glorious orange flower for a month. The blooms are past their best now, but still stand proud. The summer shade and autumn redness of the grapevine are long gone, and I have pruned it back to the twining stems around the pergola. I am tempted to pat it to sleep for a few months rest.

Inside, messmate posts hold up the house. You can see their intimate treeness in the grain exposed by the saw. The chisel marks remind us of the labour of the man who built the house. The fire keeps us warm, eating wood collected and stacked in the autumn. Eucalyptus is best, but we are also still burning pine that blew down two years ago, and silky oak that we felled because it was dead. All have been sawn to the length of the fireplace by a chainsaw powered by dead plants from millions of years ago. The wood is dry, and the fire is quick to welcome us when we arrive just on dusk. It soothes our frazzled selves.

On a cool still day we decide to burn – some bonfires piled up with rubbish cleared last summer, and the slope in front of the house to protect us from fires in the summer to come. We know the man who built the house, a park ranger, used to do this. He left us his drip torch. We have watched and recorded how Indigenous people do it in northern Australia. Things are dry enough, despite recent rain. It must have been the winds that knocked daffodil over. But we are cautious, not comfortable that fire is our friend, nor that we have the skills to do this properly. We ready the hose, the rake, the water pump. Matches made from Swedish birch trees are put to clumps of dry bracken.

It goes well. The breeze springs up at different times, directing flames this way and that, but never in difficult or unwanted places. We know the whole place needs a good burn, but we keep it inside the fence, sure that Biddy[4] would be laughing at our timidity. The fire goes where it wants to go, and some patches remain unburned. But you can see and feel the bulk of the fuel reduce on the slope. As confidence increases, we put a flame to cycad out the front, hoping it will thank us by fruiting. Not that we have the skills to hold a cycad party.[5] For a few moments, we feel in tune with the rhythms of fire, wind and planty death and renewal. Is this what it means to listen to Country? In harnessing these other forces, we are in a kind of control. But it could easily have been otherwise, if the wind had done otherwise and brought menace rather than assistance. Pleased with our work, we stop and eat plants from far away, brought here using more dead plants from millions of years ago; wheat, chickpeas, cucumbers, tomatoes. We drink tea. We have used some calories in our work, and the plants give us some more.

That night it rained, easing any fears we had about it springing up again. But the next day, still in the middle of winter, there were bushfires in the Blue Mountains.[6] Fires have been raging in the drought-ravaged western states of the USA for some weeks, and Australians watch with sympathy. What does it mean when the fire seasons in North America and Australia start to converge?

***

One of the most important things we have learned from Indigenous encounters is that they throw new light on everyday practices taken for granted in non-Indigenous worlds, including the settler ones that have shaped much of the contemporary Australian landscape. The following chapters explore some of those everyday practices and vernacular experiences. These experiences both reinforce and rupture the human/nature dichotomy in different contexts, and offer new ways to imagine more hopeful worlds.

Notes

1 Bundanon was a gift to the nation from the artistic Boyd family of their beautiful property on the Shoalhaven River in southern New South Wales. (This landscape is

depicted in many pictures by Arthur Boyd.) The property is managed as an artistic and cultural centre providing opportunities for retreats and public gatherings, such as the Siteworks event.
2 I am indebted to Lesley Instone for a discussion around this idea.
3 The term 'non-human' is of course problematic in that it defines everything against a human norm. That is the reason I mostly use the somewhat clumsy more-than-human.
4 Our teacher in Aboriginal burning practices was Murinbatha woman Biddy Simon. See Head and Fullagar (1991), Head (1994).
5 Cycads are poisonous (Beaton 1982).
6 www.smh.com.au/nsw/firefighters-battle-blue-mountains-fires-gusty-winds-hamper-efforts-20150802-gipykg. Last accessed August 2015.

References

Anderson, K. 2007. *Race and the Crisis of Humanism*. London: UCL Press.
Anderson, K. 2014. "Mind over matter? On decentring the human in Human Geography." *Cultural Geographies* 21(1): 1–15.
Anderson, K. and C. Perrin. 2015. "New materialism and the stuff of humanism." *Australian Humanities Review* 58: 1–15.
Bawaka Country, including S. Suchet-Pearson, S. Wright, K. Lloyd and L. Burarrwanga. 2013. "Caring as Country: Towards an ontology of co-becoming in natural resource management." *Asia Pacific Viewpoint* 54(2): 185–197.
Bawaka Country, including S. Wright, S. Suchet-Pearson, K. Lloyd, L. Burarrwanga, R. Ganambarr, M. Ganambarr-Stubbs, B. Ganambarr and D. Maymuru. 2015a. "Working with and learning from Country: Decentring human author-ity." *Cultural Geographies* 22(2): 269–283.
Bawaka Country, including S. Wright, S. Suchet-Pearson, K. Lloyd, L. Burarrwanga, R. Ganambarr, M. Ganambarr-Stubbs, B. Ganambarr, D. Maymuru and J. Sweeney. 2015b. "Co-becoming Bawaka: Towards a relational understanding of place/space." *Progress in Human Geography,* DOI: 10.1177/0309132515589437.
Beaton, J. M. 1982. "Fire and water: Aspects of Australian Aboriginal management of cycads." *Archaeology in Oceania* 17: 51–58.
Brumm, A., F. Aziz, G. D. van den Bergh, M. J. Morwood, M. W. Moore, I. Kurniawan, D. R. Hobbs and R. Fullagar. 2006. "Early stone technology on Flores and its implications for *Homo floresiensis.*" *Nature* 441: 624–628.
Castree, N. 2015. *Making Sense of Nature: Representation, Politics and Democracy*. Abingdon: Routledge.
Chakrabarty, D. 2012. "Postcolonial studies and the challenge of climate change." *New Literary History* 43(1): 1–18.
De Landa, M. 1997. *A Thousand Years of Nonlinear History*. New York: Zone Books.
Ellis, E. C. 2013. "Sustaining biodiversity and people in the world's anthropogenic biomes." *Current Opinion in Environmental Sustainability* 5: 368–372.
Ellis, E. C., E. C. Antill and H. Kreft. 2012. "All is not loss: Plant biodiversity in the Anthropocene." *PloS ONE* 7(1): e30535.
Firn, R. 2004. "Plant intelligence: An alternative point of view." *Annals of Botany* 93: 345–351.
Gibbons, A. 2011. "A new view of the birth of *Homo sapiens.*" *Science* 331: 392–394.
Gibson-Graham, J. K. 2008. "Diverse economies: Performative practices for 'other worlds'." *Progress in Human Geography* 32(5): 613–632.

Graham L. E., M. E. Cook, J. S. Busse. 2000. "The origin of plants: Body plan changes contributing to a major evolutionary radiation." *Proceedings of the National Academy of Sciences* 97(9): 4535–4540.

Hamilton, C. 2010. *Requiem for a Species.* Sydney: Allen and Unwin.

Hawks, J. 2013. "Significance of Neanderthal and Denisovan genomes in human evolution." *Annual Review of Anthropology* 42: 433–449.

Head, L. 1994. "Landscapes socialised by fire: Post-contact changes in Aboriginal fire use in northern Australia, and implications for prehistory." *Archaeology in Oceania* 29: 172–181.

Head, L., J. Atchison and C. Phillips. 2015. "The distinctive capacities of plants: Re-thinking difference via invasive species." *Transactions of the Institute of British Geographers* 40(3): 399–413.

Head, L. and R. Fullagar 1991. "'We all la one land.' Pastoral excisions and Aboriginal resource use." *Australian Aboriginal Studies* 1990/1991: 39–52.

Head, L. and C. Gibson. 2012. "Becoming differently modern: Geographic contributions to a generative climate politics." *Progress in Human Geography* 36(6): 699–714.

Head, L. and P. Muir. 2006. "Edges of connection: Reconceptualising the human role in urban biogeography." *Australian Geographer* 37(1): 87–101.

Head, L. and P. Muir. 2007. *Backyard: Nature and Culture in Suburban Australia.* Wollongong, NSW: University of Wollongong Press.

Hird, M. J. and C. Roberts. 2011. "Feminism theorises the nonhuman." *Feminist Theory* 12(2): 109–117.

Howitt, R. 1993. "'A world in a grain of sand': Towards a reconceptualisation of geographical scale." *Australian Geographer* 24(1): 33–44.

Howitt, R. 2001. "Frontiers, borders, edges: Liminal challenges to the hegemony of exclusion." *Australian Geographical Studies* 39(2): 233–245.

Howitt, R. 2002. "Scale and the other: Levinas and geography." *Geoforum* 33(3): 299–313.

Howitt, R. and S. Jackson. 1998. "Some things do change: Indigenous rights, geographers and geography in Australia." *Australian Geographer* 29(2): 155–173.

Howitt, R. and P. Suchet-Pearson. 2006. "Rethinking the building blocks: Ontological pluralism and the idea of 'management'." *Geografiska Annaler: Series B* 88(3): 323–335.

Instone, L. and A. Taylor. 2014. "Thinking through the Anthropocene, co-habiting with other species, co-shaping entangled futures." At the *Unnatural Futures Conference*. 3–4 July. University of Tasmania, Australia.

Jackson, M. 2014. "Composing postcolonial geographies: Postconstructivism, ecology and overcoming ontologies of critique." *Singapore Journal of Tropical Geography* 35: 72–87.

Kirksey, S. E. and S. Helmreich. 2010. "Emergence of multispecies ethnography." *Cultural Anthropology* 25(4): 545–576.

Lulka, D. 2009. "The residual humanism of hybridity: Retaining a sense of the earth." *Transactions of the Institute of British Geographers* 34(3): 378–393.

Lynas, M. 2011. *The God Species: Saving the Planet in the Age of Humans*. London: Fourth Estate.

Martin, K. 2003. "Ways of knowing, being and doing: A theoretical framework and methods for indigenous and indigenist re-search." *Journal of Australian Studies* 27(76): 203–214.

Maudlin, T. 2013 "The calibrated cosmos". Accessed 4 August 2015. Available at http://aeon.co/magazine/science/why-does-the-universe-appear-fine-tuned-for-life/.

Mitchell, T. 2002. *Rule of Experts: Egypt, Techno-Politics, Modernity*. Berkeley: University of California Press.

Palmer, L. 2015. *Water Politics and Spiritual Ecology: Custom, Environmental Governance and Development*. Abingdon: Routledge.

Plumwood, V. 2006. "The concept of a cultural landscape: Nature, culture and agency of the land." *Ethics and the Environment* 11(2): 115–150.

Roberts, R. G., K. E. Westaway, J-x. Zhao, C. S. M. Turney, M. I. Bird, W. J. Rink and L. K. Fifield. 2009. "Geochronology of cave deposits at Liang Bua and of adjacent river terraces in the Wae Racang valley, western Flores, Indonesia: A synthesis of age estimates for the type locality of *Homo floresiensis*." *Journal of Human Evolution* 57: 484–502.

Rose, D. B. 1996. *Nourishing Terrains: Australian Aboriginal Views of Landscape and Wilderness*. Canberra: Australian Heritage Commission.

Rose, D. B. 1999. "Ecological ethics from rights to recognition: Multiple spheres of justice for humans, animals and nature." In *Global Ethics and Environment*, edited by N. Low, 175–187. London: Routledge.

Rudd, H., J. Vala and V. Schaefer. 2002. "Importance of backyard habitat in a comprehensive biodiversity conservation strategy: A connectivity analysis of urban green spaces." *Restoration Ecology* 10(2): 368–375.

Stringer, C. 2012. "Evolution: What makes a modern human." *Nature* 485: 33–35.

Suchet, S. 2002: "'Totally wild'? Colonising discourses, indigenous knowledges and managing wildlife." *Australian Geographer* 33(2): 141–157.

Sundberg, J. 2014. "Decolonizing posthumanist geographies." *Cultural Geographies* 21(1): 33–47.

Taylor, M. 2015. *The Political Ecology of Climate Change Adaptation: Livelihoods, Agrarian Change and the Conflicts of Development*. Abingdon: Routledge.

Urry, J. 2011. *Climate Change and Society*. Cambridge: Polity Press.

Weir, J. K. 2009. *Murray River Country: An Ecological Dialogue with Traditional Owners*. Canberra: Aboriginal Studies Press.

Whatmore, S. 2002. *Hybrid Geographies: Natures, Cultures, Spaces*. London: Sage.

Yusoff, K. 2013. "Geologic life: Prehistory, climate, futures in the Anthropocene." *Environment and Planning D: Society and Space* 31(5): 779–795.

5 Practising hope

> *You're kind of schizophrenic in one way, you know, part of your brain thinks 'Shit, this is really bad and my kids don't have a future, or at least my grandkids don't have a future', and another part of your brain just sort of says 'Well, I've just got to keep going anyway'.*
>
> (Susan)[1]

This quote from one of Australia's leading climate scientists, interviewed in 2014, illustrates two of the dilemmas surrounding hope. First, it illustrates how scientists articulate the split between their emotional engagement with an uncertain future and their rational response to manage this emotion. Second, it exemplifies the common sense understanding that we usually associate pessimistic thoughts with negative feelings, and the converse, that we associate hope with optimism and positive feelings. In just 'keeping going anyway', Susan embodies the concept of hope as practice, a concept I aim to build through this chapter.

The chapter further develops concepts advanced by geographer Ben Anderson, who argues that although 'hope, and hoping, are taken-for-granted parts of the affective fabric of contemporary Western everyday life' (2006a, p. 733), they have not been well theorised. Anderson's depiction of hope has four particular points of relevance for us in relation to the Anthropocene.

1. Hope is understood as a process that creates possibility and potential, or at least opens up spaces in present day reality for things to be done differently.
2. It carries with it melancholy and grief.
3. It risks disappointment and has no guarantees.
4. It is everyday. This provides a starting point for how we might imagine worlds otherwise.

I build on Anderson's theorisation to decouple emotions from an understanding of hope and then to re-attach them. We specifically need to decouple the conflation of hope with optimism. For Anderson, optimism constitutes blind faith; 'Optimism, in comparison to hope, avoids encounters with the emergence of actually existing potentialities and possibilities' (2006b, p. 705). My aim is not to separate emotions and hope, but rather to recognise that a broad range of emotions, including painful ones, are entangled in hope. I argue that associating hope only with optimism acts to close down possibilities rather than open them up. I seek then to articulate a concept of hope as embodied practice that will lead us into the following chapters. Through paying explicit attention to the emotional labour involved in climate science, this chapter builds an argument that hope is something to be practised rather than felt. In so doing we consider the possibilities – and difficulties – of painful emotions being generative.

The empirical resource that I draw on in developing this understanding of hope is a research project about how climate change scientists envisage the future. Climate scientists spend a lot of time (both on and off the job) thinking about the future. As professional data crunchers, projectors and informants for planning, they have become inadvertent experts about aspects of our collective futures. If climate scientists constitute something of a canary in the coal mine for the rest of us, there is an urgent need to consider how to engage with negative emotions around climate change. What can we learn from them about hope, fear and how to respond to the challenges of climate change?[2]

These scientists feel they have to remain dispassionate, in order to be good scientists. They put pressure on themselves to separate the mind and body – to 'keep the heart a long way from the brain', as one of them put it – notwithstanding that their materialist perspective on the world would show how connected the two are within a body. This is partly, but not only, due to the wider social context of climate change being debated as a question of faith in science.

Social studies of science have only recently turned attention to the importance of emotions in scientific work (Shapin 2010, Barbalet 2011). Spencer and Walby (2013) found that, contrary to the myth of the dispassionate scientist, emotions in science were pervasive. They elaborated three specific forms of emotion, expressed in how the community shared success and failure, overcame frictions and obstacles within and between groups and negotiated non-work demands. Despite the evidence that illustrates that climate science is infused with a range of emotional tensions and relations, the myth of the dispassionate norm in science is persistent. This is further emphasised by Brysse et al., who show how it leads to scientists being 'biased not toward alarmism but rather the reverse' (2013, p. 327). Using examples from predictions of Arctic ozone depletion and the possible disintegration of the West Antarctic ice sheet, they argue that:

> the scientific values of rationality, dispassion, and self-restraint tend to lead scientists to demand greater levels of evidence in support of surprising, dramatic, or alarming conclusions than in support of conclusions that are less surprising, less alarming, or more consistent with the scientific status quo. Restraint is a community norm in science, and it tends to lead many scientists . . . to be cautious rather than alarmist, dispassionate rather than emotional, understated rather than overstated, restrained rather than excessive, and above all, moderate rather than dramatic.
>
> (2013, pp. 327–328)

Before we return to the climate scientists, we consider the dimensions of hope put forward by Anderson and others.

Conceptualising hope

Hope carries melancholy and grief

Through a study of musical affect, Anderson showed how hope is embodied; it is held in the same bodies as melancholy and grief. For one of his study participants, listening to sad music 'facilitates the possibility of him being able to "get on with it a bit"' (Anderson 2006a, p. 744). For another participant, soul music 'comes to induce and amplify a disposition of hopefulness through a disruption of the transmission of grief' (Anderson 2006a, p. 745).

Hope opens up possibility and potential

Anderson is one of a number of scholars to recognise the important connections between hope and spaces of possibility:

> Hopefulness, therefore, exemplifies a disposition that provides a dynamic imperative *to action* in that it enables bodies to *go on*. As a positive change in the passage of affect it opens the space – time that it emerges from to a renewed *feeling* of possibility.
>
> (2006a, p. 744, emphasis in original)

Anderson draws on the work of philosopher Ernst Bloch (1986). For him, Bloch's work is:

> remarkable for how it is animated by the simple principle that the existence of hope discloses how "the world itself, just as it is in a mess, is also in a state of unfinishedness and in experimental process out of that mess".
>
> (Bloch 1986, p. 221, Anderson 2006b, p. 693)

> The disposition of hope is best defined as a relation of suspension that discloses the future as open whilst enabling a seemingly paradoxical capacity to dwell more intensely in points of divergence within encounters that diminish. Becoming hopeful is therefore different from becoming optimistic.
>
> (Anderson 2006a, p. 747)

As Anderson recognises, many of the characteristics of hope he discusses are expressed in the community economies work inspired by Gibson-Graham (2006). Gibson-Graham talks about cracking open the lenses of possibility, paying particular attention to identifying and harnessing vernacular capacities and reframing to generate practical possibilities where none formerly existed (Gibson-Graham 2008, Head & Gibson 2012). Another example is Bradley and Hedrén's rethinking of how 'utopian thought can be applied in a contemporary context, as critique and in imagining and practicing desired futures' (2014, p. 2).

These possibilities and practices are also everyday ones. Anderson and Fenton (2008, p. 76) emphasise the everyday nature and spaces of hope: 'bearing witness to spaces of hope involves thinking of ordinary, quotidian, life as not-yet-become'. Hope does things differently and does many different things in these spaces; it is not a single thing. The particular local, everyday and often domestic capacities discussed in the community economies work – community gardens, sharing and trading schemes – have been much critiqued as too small, too local, too disconnected from the bigger more structural sources of economic power. But they are one space/time in which barely visible progressive futures might be rendered visible and imaginable. In later chapters we consider others.

The hope of reframing the world requires rupture

Moments of rupture or contexts of change and uncertainty are the conditions that can create the spaces for such possibilities to emerge (see also De Landa's discussion of the role of friction in non-linear dynamics (1997, pp. 41–42)). Gibson and I (Head & Gibson 2012) thought of ruptures as generative moments. These ruptures can begin as very tiny moments yet have the potential to generate unseen possibilities. As we saw in Chapter 2, Farbotko and McGregor (2010) revealed that even in formal power-laden settings such as international summits, rupture can occur through the play of emotional tensions in the form of a policy advisor's tears. This generative potential of emotion resonates with Ahmed's discussion of the role of feminist anger in 'opening up the future':

> being against something does not end with 'that which one is against'. Anger does not necessarily become 'stuck' on its object, although that object may remain sticky and compelling. Being against something is also being for something, but something that has yet to be articulated or is not yet.
>
> (2004, p. 175)

A key point here is that these moments of rupture will not all be happy moments. Their potential is driven by painful emotions. Increasingly, we need to consider that they may be violent, catastrophic and traumatic. They carry friction and all sorts of tensions and fragilities in them. As we discussed in Chapter 2, scholars are now considering how grief and violence can be considered as a resource (Yusoff 2012, van Dooren 2014). Yusoff quotes Butler (2004), on turning grief into a resource for politics. Butler was discussing post 9/11 grief in the USA, and the possibility of turning it into less hubristic responses than war (not achieved, then or since). So it is not only, as I argued earlier, that we need to carry or bear our grief with us in 'moving forward' but we need specifically to explore the way that violence and grief can potentially be utilised (Yusoff 2012). What is the performative, generative role of such emotions, and what are the constraints on that potential being realised?

Hope plays out in a non-linear world, and risks failure

Building on this idea of an uncertain future with unknown possibilities and having analysed human and other history in terms of non-linear processes of destratification and restratification, De Landa contemplated the possibility of rupture. He described what he calls the necessary 'destratification' of present day reality:

> Changing our way of thinking about the world is a necessary first step, but it is by no means sufficient: we will need to *destratify reality itself*, and we must do so without the guarantee of a golden age ahead, knowing full well the dangers and possible restratifications we may face.... There is, indeed, a new kind of hope implicit in these new views. After all, many of the most beautiful and inspiring things on our planet may have been created through destratification. A good example of this may be the emergence of birdsongs: the mouth became destratified when it ceased to be a strictly alimentary organ, caught up in the day-to-day eating of flesh, and began to generate other flows (memes) and structures (songs) where the meshwork element dominated the hierarchical.
>
> (1997, p. 274)

Though possibly catastrophic, destratification or decoupling from an embedded idea, action or practice may give rise to novel and unthought-of restratifications. For De Landa this process of facing restratification is not to be confused with despair or nihilism. On the contrary it offers possibilities, in fact it may force us to rework how we think and act.

Clearly this is an understanding of hope that, while it contains considerable possibility, has no guarantee of success. Nor does it even aim towards an ideal of progress: 'There can therefore be no teleology, or guarantee, in a world where hope acts as a principle of "unfinishedness, of only fragmentary essential being in all objects"' (Bloch 1986, p. 221, quoted in Anderson 2006b, p. 693). And it will be messy. As Urry (2011) makes clear in his discussion of shifts towards a low carbon future, system-wide change can happen in ways that are forced from above, that are generated by widespread suffering, or that might see the emergence of regional warlords. The story is not necessarily a happy one. When we narrate it afterwards, of course, this history often looks quite different. As does the Anthropocene. We do not think often enough that James Watt was just experimenting with his steam engine, or that hunter-gatherers were experimenting different plant practices in the emergence of agriculture.

Hope as practice and experiment

Hope has an aspect of being able to 'go on', as seen among Anderson's interviewees, and as we will see in the examples of climate scientists below. For them, hope is action. This notion of hope as practice has characteristics in common with Annemarie Mol's (2006) concept of 'tinkering'. It is inherently experimental.

Mol looked at treatments for patients with atherosclerosis of the leg vessels, specifically a comparison between two treatments, but not in a context where one treatment was clearly more effective than the other. 'Instead, both therapies had a positive effect, but these effects were different' (Mol 2006, p. 407). What, then, is good care or, in this case, the better treatment? Her answer was that it depends. But not on more scientific trials, or sorting out the messiness of the social context:

> The good, in this case, is not unequivocal. There are different goods at play, and somehow this complication has to be lived and dealt with. Note that this is not a merely social matter. It is not the case that the body, and its disease, is singular and easy to survey, while as soon as we attend to illness, to social life, as well as to a disease, things become messy and complex ... the body itself is as implicated in what is happening as one's social life.
>
> (Mol 2006, p. 407)

Mol uses verbs like loving, tinkering, doctoring, caring, letting go, as replacements for 'act'. She uses tinkering as being:

> Less strategic in its connotation than "co-ordination", and better at stressing an ongoing effort than "association" ... This suggests persistent activity done bit by bit, one step after another, without an overall plan. Cathedrals have been built in a tinkering mode, and signallers or aircraft designers also work in this way.
>
> (2010, pp. 264–265)

This is very different to modernist terms we are more familiar with, such as manage, intervene, control.

Coombes et al. carry an experimental structure of feeling in their argument that Indigenous examples provide prototypes for the geographies of hope:

> it is precisely because those possibilities are born of contradiction, messiness and the influence of the past that the oracular visions of postcolonial theory cannot be precise, and that the emancipatory projects which it foreshadows are often compromised, fragile or contested.
>
> (2013, p. 692)

In response to critique that dismisses Indigenous situations as naive and localist, they make a particular point of the experimental potential of apparently unlikely juxtapositions. Foremost among these is the emancipatory potential of the Indigenous as urban (Pieris 2012). James (2012) discusses the potential of inter-cultural experiments, including both Indigenous and migrant groups. While definitely not without friction, these open up new possibilities for how we might imagine the urban.

In summary then, the hope that emerges from this review is fragile and messy. It is a carrier of grief, in tension with many things including the possibility of failure. It is materially embodied, urgent, situated, grounded and realistic. In practising interventions in which causes are not clear or linear, it needs to be open to surprise and the unexpected. It is consistent with what Gibson-Graham (2008) called a 'performative ontological project'.

Climate change scientists and the future – keeping the heart a long way from the brain

In this section I draw on the results from an empirical study conducted in 2014 (Head & Harada in prep.). A sample of Australia's leading climate scientists took part in interviews which explored how they viewed the future.[3] In this discussion, I am treating optimism and pessimism as dispositions towards the future that contain mixtures of emotions. Optimism tends to contain more pleasant emotions; excitement, happiness, anticipation. Pessimism tends to contain more painful ones; anxiety, grief, fear.

The optimism/pessimism continuum

When asked to place themselves on a continuum between optimism and pessimism in their feelings about climate change, there was a broad spread of answers, often from the same person. Common answers included 'It depends on the day', with people describing their feelings as 'wildly oscillatory', 'schizophrenic' and 'mixed'. Optimism was a 'defensive mechanism to keep yourself motivated' and 'compulsory'. No one was unambiguously optimistic, and there were many more expressions of pessimistic than optimistic thought.

In characterising themselves as either pessimists or optimists, many referred in their answers to the scientific evidence, and to probabilities or quantitative risks of certain things happening. Yet even so, Ed points to the emotional conflict that this causes, finding it a difficult question to answer 'rationally'. For example, he moves from numbers to emotions in this answer:

> I guess scientists including me sometimes retreat to numbers when asked about optimism versus pessimism, just let me begin with numbers. I think 2 degrees [of warming] is off the table. Two and a half is maybe still on the table. Three and four are looking the more likely outcomes, based on a rough guess as to how fast action can be taken. Does that cause pessimism? Certainly it does. If, and I find that a difficult question to answer kind of rationally, because I don't know much intuitively about what a world at two and a half degrees and/or a world at four degrees would actually look like, except for the vague sense that these are . . . progressively more difficult worlds as that number creeps up.
>
> (Ed)

When asked how they imagined the future would look in an everyday sense, a common answer was that technological solutions exist, but the community, industry and government were just not pursuing them. Where optimism was found, it was almost always in relation to the possibility of technological solutions and human ingenuity, rather than political change. A number of people mentioned the widespread uptake of renewable energy, particularly community uptake of solar photovoltaic systems. Although most discussed this in technological terms, it was in fact a socio-technological change they were referring to, as it was usually couched in terms of 'bottom up', 'household' or 'community' action, as distinct from 'government' action. Some referred to this as a social 'tipping point' in the same way that there are tipping points in the earth system. While there remained an underlying belief in science as having the answers, there was widespread frustration that people, especially governments, were not 'taking notice' of the science. Thus the anger or frustration that resulted from overlooking the science can be viewed as generating possibilities for hope that solutions might arise in unexpected ways and places.

For some, there had been an increase in pessimism over the last few years as a problem that could have been quite tractable if addressed early had continued to be ignored by policy-makers:

> But yeah it's becoming increasingly – just gradually obvious [that] we've left it too late to deal with it in a sensible and cost effective way. So it's now going to cost us a lot more to deal with it than it would if we'd done it 15 years ago. So it's just a gradual sort of move towards pessimism.
>
> (Martin)

The strongest pessimism was among those who discussed the interaction between climate change and other issues, including population growth, food insecurity, ecological degradation and geopolitical conflict. For example, John said:

> So if you have three times more people than you currently do, and we're already water stressed in times of drought in our major cities, essentially that just means . . . you'll have to cut down your drought water allocation of 120 litres to something like 40 litres a day per person, which is Third World country levels . . . I just don't think people actually understand this.

Virtually all participants described their pessimism about climate change as at odds with their normal personality. Several described themselves as 'glass half full' people, and only one as a 'born pessimist'.

Distancing from emotions

Several types or expressions of distancing rationality from emotions were evident in the interviews (Table 5.1). The distancing is also for different reasons, but these scientists have in common that they feel they need to distance themselves from

Table 5.1 The emotional labour of distancing by climate scientists

Type of distancing	Examples
Being a dispassionate scientist, in order to be a good one	But getting that science right is the really hard thing and among scientists the debates are often intense on issues, of all sorts of issues around climate change, and this is a natural part of science. It's a very good and healthy part of science but it can be very intense and demanding and it doesn't stop at 5 o'clock. (Ed)
	I would prefer to describe what constitutes a catastrophe rather dispassionately, without taking on the emotional impact of that. (Gary)
Dealing with denialists	Almost all of the fundamental science that gives us clear evidence for global warming doesn't come from climate science. It comes from chemistry, physics, mathematics, statistics and yet we're held to account for failing to be able to articulately explain the relationship between quantum theory, the rotation of a CO_2 molecule and its ability to absorb infrared radiation . . . The physicists do not have to deal with deep scepticism on whether the universe is expanding. The aeronautical engineers do not have to defend whether the shape of a wing generates uplift. (Geoff)
Putting distance between work and the rest of life	I go home, that's home. It's not work . . . I want to read trashy novels and have cups of tea. (Karen)
	I can't do it 100 per cent of the time . . . [I] read detective novels and cook and garden and that sort of stuff. (Susan)
	I do not spend my weekends, nose in a book, in the science . . . (Geoff)
Preserving energy for the things that matter	I've never blogged. Lots of my colleagues got into blogging for a while and then stopped because it was just so vile. I never even started, I just see it as a place that's bad for me psychologically and that will do no good. (Brian)
	I am not going to get involved in an ongoing thing where [a sceptic] wants me to respond to every single one of his arguments. So I will be very clear about that because you can't. It would drive me potty. (Linda)
Protecting the psyche from the subject matter itself	I actually deliberately block it out as much as I can [when away from work]. So I try, I really try not to because I don't think I'd be in very good shape if I let myself think about it all the time. So there's a denial for you [laughing] . . . I have to say to be brutally honest I tend not to look at [the future] too much because it's very confronting. I have kids . . . (Linda)
Don't tell the kids	I've never talked about environmental problems with my kids at all but they hear it everywhere, I mean they know what I do. (Brian)
	I certainly worry about their future a great deal. I try not to share that with them. (Susan)

Type of distancing	Examples
It won't affect me personally	My strategy for my kids – and it's a little bit confronting – two things; a bloody good education and European passports. We have dual nationality. My kids can go and, can see Australia, can go to New Zealand, they can go to anywhere in Europe. That's a pretty good contingency plan for climate change and whatever happens in any given country . . . Now, that isn't a strategy that everybody can embark upon, but when it comes to my kids I'll be selfish thanks very much. (Geoff)
	I know people from the [research organisation] who've got their apocalypse houses down in Tasmania. You know, they've already bought property in places [which] they think aren't going to warm as much as other parts of Australia . . . [they] . . . have literally got their . . . climate change houses tucked away. (Tanya)

their emotions around the future. This distancing takes work; it constitutes emotional labour.[4] The first expression of this was the importance of being a dispassionate scientist, in order to be a good one. In a variety of ways all the interviewees spoke of the crucial importance of doing excellent science as the underpinning of everything else their work entails, and as the basis of their professional reputations. For most, this was expressed in the traditional terms of objectivity, involving a necessary distancing from emotions and from persons. Geoff described this as 'keeping the heart a long way from the brain', and Ed spoke of the possibility of being wrong as the only thing that kept him awake at night.

The importance of maintaining their objectivity as scientists is exacerbated in the face of a strong climate change denialist movement:

> Prominent people are accusing you of being a fraud . . . that's going to weigh on you . . . And we used to get lots of hate mail and death threats; that seems to have stopped a year or two ago . . . it's not something that other scientists have to deal with.
>
> (Brian)

The emotional labour of doing the work of climate science is clearly evident in this quote. The need to deal with the 'hate mail' and 'death threats' is seen as part of the job. Further, climate science bears an additional burden; the onus to 'defend' their science. As Geoff's quote (Table 5.1) makes clear, climate scientists are being held to public account in ways that other scientists are not.

Putting some distance between work and the rest of their lives was done with varying degrees of success. Switching off from work was a common strategy, yet at times scientists also described being always 'on' and never being able to separate from it. Thus, the importance of maintaining personal health, through exercise and diet, was described not just for its own sake, but also so that they could continue to do their job well.

Participants described quite emphatically decisions they have taken to preserve their energies for the things that matter. This involved being very selective in when they engage in discussions around climate change. For example, Susan said, 'I don't talk to taxi drivers . . . I love a silent taxi driver', Brian avoids the blogosphere and Linda will not argue in detail with sceptics. The need to save energy for when it matters is evident in the way some avoid discussing climate change in social situations, when they would 'rather talk about the cricket' (Nick). 'Oh God I never tell them [in a social situation, that I'm a climate scientist], it usually descends into them shouting at you, or yeah, like telling you either you're not doing enough or you're doing too much' (Karen).

Again, the influence of denialists is evident in the need for these measures, and the sense that such encounters are inevitably bruising. One (admittedly, the self-described chronic pessimist mentioned above) blamed the 'relentless attacks' from denialists, the media and politicians for creating a negative atmosphere in the whole area of work:

> you can't have kept on doing this for a long time without having a bit of . . . graveyard humour . . . If you let it get you down then you're gone, you're no longer with us. The ones that have remained are the ones who are sufficiently stoic about this to laugh off the regular attacks . . . If that sort of thing gets you down then you're no longer a climate scientist, you've moved onto something less continuously negative I guess.
>
> (Martin)

Only one participant (described by colleagues as a 'warrior' and 'gladiator', and self-described as having 'a thick skin') discussed actively monitoring climate change sceptic blogs in order to be prepared for arguments likely to come up in interviews or public talks.

It is notable that in the examples above, the things that they need to distance themselves from are a heavy and complex workload, various workplace issues and a fraught public sphere, rather than the subject matter of climate change itself. Several people described ways in which their physical and mental health had been adversely affected, to the extent of needing professional help. But even when medical help was required, it was usually for a combination of the same workplace stresses and adversarial public issues, and the need to avoid 'burnout', rather than distress about the state of the world alone. This exemplifies Whittle et al.'s (2012) argument that much emotional work goes unnoticed or undetected.

Beyond the above examples, a further type of distancing involved protecting the psyche from the subject matter of climate change itself. A number of participants recognised within themselves or their colleagues a distancing, or 'dissonance', from the evidence at hand, because of the severity of the implications:

if I'm staying awake at night because of something to do with the topic, it's often much more to do with institutional or organisational frustrations – you know, things that are going wrong in the workplace or could be better in the workplace with restructurings and things like that – than it would be about climate change per se. And that doesn't at all diminish the importance of the issue, it's just that you've got to have some way of protecting your psyche against unwanted despair and intrusion.

(John)

Here John, who had described above the possibility of Third World levels of water availability in Australia, recognises that the implications are so severe that to let them keep him awake at night would be totally disabling. Susan compared it to the way we think about death: 'we all know we're going to die but most of us who are reasonably socially well-adjusted don't think about it every day because it's paralysing, and denialism is a form of protecting people against paralysis I think' (Susan).

As Linda's explicit reference to denial (Table 5.1) illustrates, these examples constitute 'everyday denial', in the sense elaborated by Norgaard (2011) that we discussed in Chapter 2. Linda's mention of children shows a further resonance with Norgaard's findings. The reticence to discuss climate change issues with children arose a number of times. Actual responses varied according to the age and personality of the children, but all the parents who mentioned this were worried about leaving children and teenagers distressed and disempowered.

Recognising that they come from a section of society that has strong economic and social resources, participants were aware of the ways they, as individuals, would likely be buffered from the main impacts of climate change. Several had considered explicit strategies for themselves and their families, as Geoff's dual passport example demonstrates (Table 5.1).

Downplaying doom and gloom

This theme was expressed in several different ways. Although the challenges posed by climate change were seen as extremely severe, most participants felt that a doom and gloom response was the wrong one. For some this involved contrasting themselves with people outside the climate science community, who were felt to be more pessimistic than those inside. First, several felt it was inconsistent with the science:

I'm not a catastrophist. I don't see it as the end of the world, I don't see it as the worst problem. I see it as a problem we would have been able to handle quite easily starting 15, 20 years ago. We can still handle it but it will be more challenging now.

(Martin)

So I sort of don't do the doom and gloom stuff that some of the groups do . . . It's because it doesn't represent the science.

(John)

Second, it would fail to get traction with the general public, leading to paralysis and lack of action. This has parallels with the way they described the emotional work of having to keep their personal pessimism at a distance, in order to keep going, as expressed by 'compulsory optimist' Ed:

> So you kind of have to be optimistic because if you fall into at least acting from a state of pessimism then you can't break out, you get into a downward spiral. It becomes a sort of, I guess at first a personal weight, but also a weight that gets transmitted from one person to another and therefore, stops us from doing rational things on climate change. And to the extent that we need to do rational things on climate change, it's really important that we're not depressed collectively about the future. That if the future becomes something that's too big to handle and outside our control then we've lost it and so, that's why I sometimes say I'm a compulsory optimist.
>
> (Ed)

> I don't think it's actually that valuable telling people we're all doomed, because then they don't, they just feel like, 'Then what's the point of doing anything? . . . if it's only going in one direction and we can't do anything about it then I might as well just, you know, crank up the air con and sit back and enjoy myself.' . . . I don't think it's a very useful attitude to have.
>
> (Karen)

As Ed's quote above implies, this view applied even when the individual scientist was particularly pessimistic:

> once you become convinced that the world, the way it's going, is doomed, the human race is doomed, I don't see how you can do anything other than keep trying and work in this area. I don't understand why all scientists aren't working on climate change. I honestly don't, because if you don't have a habitable planet, it doesn't matter . . . nothing else matters.
>
> (Susan)

A broader societal expectation to be positive rather than negative is encapsulated in Geoff's experience:

> I have been told time and time again that one should emphasise the positives. We can do it. We can. It'll be fine, uh we just have to do this and this and this, and I think it's, like, naive . . . As an evidence based scientist I see no evidence whatsoever that the world will get to grips with this problem before it's too late, and I think it's too late to avoid dangerous climate change.
>
> (Geoff)

A third and slightly different expression of this theme was frequent distancing from two 'extremities', the denialists (who say there is nothing to worry about) and the 'doom and gloom' Green activists. This view of science as representing balance is expressed by Geoff: 'our lot critique scientifically, legitimately and that always cuts across either the sceptics' agenda or the Green advocacy agenda, because the science is always in between those two extremes'. Here Geoff wants to maintain a balance, despite in the previous quote having presented a rational explication of why it is naive to emphasise the positives. This is an example of 'Erring on the Side of Least Drama', as expressed by Brysse et al. (2013), a point I return to below.

Emotions of engagement and empowerment

It is important to acknowledge that this sample comprised more or less-resilient survivors of a competitive academic system. Nevertheless, although they are demonstrably more pessimistic than optimistic, as outlined above, none had given up. They were continuing to practise their science and the associated set of mentoring, capacity-building, advocacy, policy, institutional-labour and public-engagement activities. Nearly everyone expressed a need to 'keep on going' because the alternative was too awful to contemplate, and also because most of them loved their jobs. The strongest emotion was expressed when people described their work, and described science as a particular source of pleasure, being 'cool', 'interesting' and fun:

> I got involved primarily in a science perspective on climate change because I felt that it was a really interesting science problem.
>
> (Thomas)

> [T]he science is so intriguing, so interesting because it's such a vital issue and it's . . . fascinating. So when you take your human – sort of, I'm a human and I've got kids and they need to have a future and – take that hat off, you can get really fascinated with how this planet's changing and how we're pushing it.
>
> (Nick)

> So you need to be able to make it simple [for the public] and stay faithful to the science. And that's not easy. I actually find that a very interesting . . . intellectual challenge, to be able to do that.
>
> (Nick)

The 'keeping on going' was expressed as:

> We've just got to get on with it . . . The government could be helping a lot more, but actually we've got to get on with it anyway.
>
> (Paul)

We really are screwed if we all give up.

(Susan)

Or, as Ed said above, 'to the extent that we need to do rational things on climate change, it's really important that we're not depressed collectively about the future'. While acknowledging the problems they faced in trying to effect change, their practices of endurance embody hope as an everyday practice.

Painful emotions, hope and generative change?

In considering how we can weave the conceptualisation of hope with insights from the scientists' interviews, my point is not to create more burdens and responsibilities for a group of dedicated professionals and citizens. Nor am I suggesting that society should look to climate scientists for particular insights into how to behave, or how to prepare for the future. If hope is necessarily messy, experimental and partially constituted in difference, it explicitly should not take any particular empirical study as a guide or a source. I do use this empirical example to show some of the ways in which performing a new experimental view of hope is not as easy as it might seem. In exposing some of the emotional labour involved in climate science, we can consider our own emotional labour, and see how current understandings of hope remain too fixed on a particular kind of cheerfulness.

The emotional dimensions of working with climate change data were both pleasurable and painful, shifting on a daily basis to evoke moods of both optimism and pessimism. As described above, these scientists experience mixtures of optimism and pessimism in their response to climate change and its implications for likely futures, but their collective pessimism, influenced by their interpretation of the scientific evidence, outweighs their optimism. For most of them, that pessimism is distinct from the way they would describe their underlying personality.

The multiple ways that the scientists attempt to distance themselves from emotions around their work can be understood as constituting a kind of everyday denial, as conceptualised by Norgaard (2011). Norgaard illustrates how this kind of denial is socially organised and reproduced, arguing that 'it *takes work* to ignore the proverbial elephant in the room' (2011, p. 93, emphasis added). In this case the elephant is painful emotions such as despair, and we have seen in the interview analysis different manifestations of the labour involved in keeping the elephant at a distance. Several factors are involved: the importance of upholding the identity and reputation of the objective scientist, particularly against a relentless climate change denialist attack; the need to maintain personal wellbeing in demanding professional lives; and a protection of the personal psyche against worst-case scenarios, that some participants referred to as denial. There was an uncanny resonance here in the reluctance in both studies to discuss climate change too much with children. The scientists are angry at society's failure to take notice, but they have their own kind of distancing from the implications of their work.

Notwithstanding increasing analysis of the role of emotion in scientific endeavour, these scientists work in a context where the predominant view of science continues to be that the heart (emotions) is kept a long way from the brain (rationality). Or more specifically, certain kinds of emotions are permissible. They attempt to protect themselves from stress and anxiety by downplaying what they understand as inappropriate emotions and inducing the positive ones (love of their job, passion for science). Emotions expressed as 'love' or 'fascination' for their work also worked to stimulate engagement and connection, and a resolute commitment to keep going. In this respect they illustrate the concept of hope as practice. There was a lot of laughter in the interviews, and a number of people talked about how they valued the collegial relationships of their workplaces and of the broader climate science community. As noted above, the sample is already biased towards personal resilience and the ability to survive and succeed in a competitive academic environment of limited resources. The number of opportunities shrink at each career stage, creating a funnelling effect. Even the early career participants here are already survivors of such a system, although they are concerned about how long they will be able to be effective. Although I have no medical expertise, it is hard not to agree with Fritze et al.'s (2008, p. 7) argument that people working in climate change may be at particular risk of mental health issues. As we saw in Chapter 2, scholars in the social dimensions of climate change and disaster research are starting to write more explicitly about these issues (see also Willis 2012). For social scientists, self-reflexivity is a standard research tool, so it is not surprising to see this, albeit it still takes professional courage to expose one's own vulnerability. It is less common in the natural or physical sciences, where there is not a research language to discuss such things. In this respect feminist approaches that attend to the productive relations between embodied emotions and rationality can provide productive tools for science.

This double move of downplaying painful emotions and celebrating or fostering pleasurable ones enables the scientists to keep going, and we can all identify with this. None of us are built to welcome unpleasant things! But we can also see here how it leads to an unjustified bias towards positive scenarios (Anderson & Bows 2012). By rationalising emotions they systematically avoid worst-case scenarios, even where they are as statistically likely to occur as not. A variety of processes contribute to downplaying potentially severe consequences, consistent with Brysse et al.'s (2013) argument for Erring on the Side of Least Drama. In the context of the inherent conservatism of the scientific method, in which 'generally, the burden of proof is on those who wish to change prevailing views and approaches' (p. 334), Brysse et al. (2013) suggest two reasons for ESLD. First, 'the historic link between uniformitarianism, anti-clericalism, and the rise of modern geology and evolutionary biology', and second, 'that the basic, core values of scientific rationality contribute to an unintended bias against dramatic outcomes' (p. 334). Several aspects of our findings resonate here, including the distancing from catastrophist thought ('doom and gloom') among the general public and Green advocates, and the 'compulsory optimism' of keeping going regardless.

The supposedly rational and dispassionate context of climate change analysis could be expected to have a particularly complex interplay of emotions. Nevertheless the study provokes important questions for all of us in a context of strong social and cultural pressure to be positive rather than negative. Rational, embodied responses suggest we will need to bear painful emotions (fear, grief, anxiety) if we are to be effective and truthful. It is clear that the scientists have not yet found a way for their painful emotions to be generative of new possibilities, to contribute to the social transformations they argue are needed. Indeed they can barely express such thoughts. As a society, we can work hard to help one another not be depressed, but we urgently need to find ways for society to bear, and bear witness to, the painful emotions around climate change. But can we go further? We need to consider, in a profound cultural and psychological sense, why painful or difficult emotions are paralysing, and whether there are ways to carry such emotions differently to energise action. We do not yet have good models for these kinds of emotions to be generative (Willis 2012), and it is a major task ahead of us.

Drawing on a range of researchers, and on the demonstrated practice of the climate scientists, we can see one expression of hope as practice, as keeping on going. So, the hope in this chapter is not an optimistic affect, nor a utopian dream, nor a sunny disposition. Hope is practised and performed; it is a sort of hybrid, vernacular collective worked out in everyday practice and experience. It amplifies and inverts some of the things we are already doing. In the following chapters, I explore some different potential expressions of hope as practised.

Notes

1 Pseudonyms are used throughout this chapter.
2 This topic is of broad interest. A number of journalists have interviewed climate scientists about their emotional responses to the future, as in this example titled 'When the end of human civilization is your day job' (Richardson 2015). At the Melbourne Writers Festival in August 2015 an exhibition of twenty-two handwritten letters from some of Australia's leading climate researchers describes how climate change makes them feel (Melbourne Writers Festival, 2015).
3 For full details of this study see Head and Harada (in prep.). The sample comprised four females and nine males. Ages ranged from the early career stage (defined in Australia as within five years of PhD completion) to retired but still academically active. At the time of interview they were employed in, or had adjunct appointments at, universities and/or government scientific bodies. Interviews took place in or close to participant workplaces in eastern Australia between June and October 2014. The term 'climate scientists' as used here encompasses two main areas of expertise. Some interviewees were atmospheric climate scientists with disciplinary backgrounds in physics and maths. Others have more general backgrounds in the earth, environmental and biological sciences, and tend to work more on the implications of climate change. In the quotes used here, pseudonyms are used for individual responses that exemplify wider trends. In order to preserve anonymity, more detailed demographic descriptors are not used.
4 Some authors distinguish between emotional labour that occurs in a paid-work context, and emotional work that takes place in a home or unpaid context (Whittle et al. 2012). I consider that to be a spurious distinction in this discussion, since the emotional labour so clearly crosses the work/home boundary. I am more interested in identifying the types of

labour/work that are occurring. The terms 'emotional labour' and 'emotional work' are used interchangeably in this discussion.

References

Ahmed, S. 2004. *The Cultural Politics of Emotion*. New York: Routledge.
Anderson, B. 2006a. "Becoming and being hopeful: Towards a theory of affect." *Environment and Planning D: Society and Space* 24(5): 733–752.
Anderson, B. 2006b. "'Transcending without transcendence': Utopianism and an ethos of hope." *Antipode* 38(4): 691–710.
Anderson, B. and J. Fenton. 2008. "Editorial introduction: Spaces of hope." *Space and Culture* 11(2): 76–80.
Anderson, K. and A. Bows. 2012. "A new paradigm of climate change." *Nature Climate Change* 2: 639–640.
Barbalet, J., 2011. "Emotions beyond regulation: Backgrounded emotions in science and trust." *Emotion Review* 3: 36–43.
Bloch, E. 1986. *The Principles of Hope*, translated by N. Plaice, S. Plaice and P. Knight. Cambridge, MA: MIT Press.
Bradley, K. and J. Hedrén. 2014. "Utopian thought in the making of green futures." In *Green Utopianism: Perspectives, Politics and Micro-Practices*, edited by K. Bradley and J. Hedrén, 1–20. New York: Routledge.
Brysse, K., N. Oreskes, J. O'Reilly and M. Oppenheimer. 2013. "Climate change prediction: Erring on the side of least drama?" *Global Environmental Change* 23: 327–337.
Butler, J. 2004. *Precarious Life: The Powers of Mourning and Violence*. London: Verso.
Coombes, B., J. T. Johnson and R. Howitt. 2013. "Indigenous geographies II: The aspirational spaces in postcolonial politics – reconciliation, belonging and social provision." *Progress in Human Geography* 37(5): 691–700.
De Landa, M. 1997. *A Thousand Years of Nonlinear History*. New York: Zone Books.
Farbotko, C. and H. McGregor. 2010. "Copenhagen, climate science and the emotional geographies of climate change." *Australian Geographer* 41(2): 159–166.
Fritze, J. G., G. A. Blashki, S. Burke and J. Wiseman 2008. "Hope, despair and transformation: Climate change and the promotion of mental health and wellbeing." *International Journal of Mental Health Systems* 2: 13.
Gibson, C., C. Farbotko, N. Gill, L. Head and G. Waitt. 2013. *Household Sustainability: Challenges and Dilemmas in Everyday Life.* Cheltenham: Edward Elgar.
Gibson-Graham, J. K. 2006. *A Postcapitalist Politics*. Minneapolis: University of Minnesota Press.
Gibson-Graham, J. K. 2008. "Diverse economies: Performative practices for 'other worlds'." *Progress in Human Geography* 32(5): 613–632.
Head, L. and C. Gibson. 2012. "Becoming differently modern: Geographic contributions to a generative climate politics." *Progress in Human Geography* 36(6): 699–714.
Head, L. and T. Harada. in prep. "Keeping the heart a long way from the brain: The emotional labour of climate scientists."
Instone, L. and A. Taylor. 2014 "Thinking through the Anthropocene, co-habiting with other species, co-shaping entangled futures." At the *Unnatural Futures Conference*. 3–4 July. University of Tasmania, Australia.
James, S. W. 2012. "Indigeneity and the intercultural city." *Postcolonial Studies* 15(2): 249–265.

Melbourne Writers Festival. 2015. "Exhibition: Is this how you feel?" Accessed 4 August 2015. Available at http://mwf.com.au/session/exhibition-is-this-how-you-feel-4/.

Mol, A. 2006. "Proving or improving: On health care research as a form of self-reflection." *Qualitative Health Research* 16(3): 405–414.

Mol, A. 2010. "Actor-Network Theory: Sensitive terms and enduring tensions." *Kölner Zeitschrift für Soziologie und Sozialpsychologie* 50(1): 253–269.

Norgaard, K. M. 2011. *Living in Denial: Climate Change, Emotions, and Everyday Life*. Cambridge, MA: MIT Press.

Pieris, A. 2012. "Occupying the centre: Indigenous presence in the Australian capital city." *Postcolonial Studies* 15(2): 221–248.

Richardson, J. H. 2015. "When the end of human civilization is your day job." *Esquire*. Accessed 4 August 2015. Available at www.esquire.com/news-politics/a36228/ballad-of-the-sad-climatologists-0815/.

Roelvink, G. 2010. "Collective action and the politics of affect." *Emotion, Space and Society* 3(2): 111–118.

Shapin, S. 2010. *Never Pure*. Baltimore, MD: Johns Hopkins University Press.

Spencer, D. C. and K. Walby. 2013. "Neo-tribalism, epistemic cultures, and the emotions of scientific knowledge construction." *Emotion, Space and Society* 7: 54–61.

Urry, J. 2011. *Climate Change and Society*. Cambridge: Polity Press.

van Dooren, T. 2014. *Flight Ways: Life and Loss at the Edge of Extinction*. New York: Columbia University Press.

Whittle, R., M. Walker, W. Medd and M. Mort. 2012. "Flood of emotions: Emotional work and long-term disaster recovery." *Emotion, Space and Society* 5: 60–69.

Willis, A. 2012. "Constructing a story to live by: Ethics, emotions and academic practice in the context of climate change." *Emotion, Space and Society* 5: 52–59.

Yusoff, K. 2012. "Aesthetics of loss: Biodiversity, banal violence and biotic subjects." *Transactions of the Institute of British Geographers* 37(4): 578–592.

6 Rethinking agriculture, rethinking Anthropocene

Typha, the bulrush, is a widespread but scarcely noticed plant in the city.[1] It makes itself at home in the damp cracks of the urban landscape – in drains, creeks and pondages along railway lines. In Wollongong you see it from the train, when you drive across a creek, and at the edges of both remnant and newly constructed wetlands (Figure 6.1). It grows close enough to our university campus for students to harvest, process and bake the starchy rhizomes in the course of a two-hour practical class, to learn about its role as a staple Aboriginal plant food (Gott 1999). (Or they did, until I decided that the polluted runoff the rhizomes were absorbing stretched my duty of care for my students beyond safe limits). The ethnographic record shows *Typha* use throughout southeastern and southwestern Australia. One record on the Murrumbidgee River 'described very large earth ovens holding "half a ton" of rhizomes, prepared for large gatherings of people' (Gott 1999, p. 40). It is likely to have been a particular staple on this and the Lachlan

Figure 6.1 *Typha* in roadside drain, Wollongong, NSW. Both old and new growth are visible. Photo: author.

River, and along the middle and lower reaches of the Murray, where it contributed to relatively high population densities. *Typha* thrives in these seasonally variable areas; it 'requires regular flooding, but survives seasonal dryness and short yearly periods of drought' (Gott 1999, p. 43). Almost as important in some areas was its use for string. Fishing and duck nets up to 50–60 metres long were made of this string, facilitating access to protein sources, in addition to the carbohydrates provided by the rhizomes themselves.[2]

Knowing this history, I have long been fascinated by *Typha*'s quiet but useful presence in our urban landscapes, where it seems to be mopping up polluted patches of water. In many rural areas it is considered a weed because of its tendency to colonise farm dams and irrigation channels, reducing the amount of open water. In my survivalist moments, I wonder whether *Typha*, the resilient survivor of colonisation and urbanisation, constitutes the archetypal Anthropocene crop – opportunistic, strong and able to thrive in diverse conditions.

The Anthropocene is still agricultural

In the geological sequence, the Anthropocene is argued to follow the Holocene, the last 11,700 years. The defining characteristics of the Holocene are warmer and wetter conditions than prevailed in the previous ice ages, and the appearance and spread of agriculture. The Holocene is often cast as a benign, stable period, providing the relatively predictable seasonal rhythm on which agricultural cycles depend. In turn, the accumulated surpluses, possible under agricultural lifeways, underpinned concentrations of humans into cities and the flowering of all that we associate with civilisation.

Does that mean the Anthropocene is somehow a period beyond agriculture? In linear trajectories of human progress, the agricultural age (or even revolution) is followed by an industrial one, a technological one in the mid-twentieth century, and the internet age of the early twenty-first century. We think of ourselves as industrial or postindustrial societies, but our subsistence base continues to be as underpinned by agriculture as it has ever been. Our urbanity is totally dependent on agriculture, yet the majority of urban populations are both spatially and mentally distant from their sources of food. Visions of the future that emphasise ongoing global urbanisation and the 'fortress city', beyond the walls of which nature is somehow considered better protected, usually neglect to discuss the footprint areas from which the source of the city's subsistence will come. Such cities are also vulnerable to catastrophes – of crop failure, of transport collapse, of market logistics. However we imagine the Anthropocene, it is not somehow post-agricultural.

Indeed, as we saw in Chapter 1, one of the candidate timings for the Anthropocene (admittedly a minority one) is agricultural (Ruddiman 2003). Regardless of whether that candidate period for the commencement of the Anthropocene prevails, the evidence shows many ways in which agriculture has transformed the surface and processes of the earth. Anthromes (anthropogenic biomes) as defined by Ellis and Ramankutty (2008), and discussed further by Ellis, were 'emerging along with agriculture more than 8000 years ago' (2013, p. 368). Anthromes revise biomes – a key analytical unit of conventional biogeography – to explicitly include human agency in vegetation systems.

In this chapter I explore the ways the Anthropocene is entwined, both conceptually and materially, with agriculture. I disentangle some of the threads that connect agriculture/Holocene/Anthropocene, in order to rethread them differently. The challenges of the Anthropocene require us to rethink not only agriculture, but also the history of how we have thought about agriculture.

But how are we are going to reframe agriculture in a non-linear world? A four degree warmer world would make the present scale of agricultural production impossible, and even a two degree warmer world will severely challenge it in many regions. But what would it do to our agricultural selves and cultures? If the Anthropocene is characterised by volatile, unpredictable and rapid change, what does that mean for the relative certainty and seasonal rhythm of agriculture? What kinds of subsistence will be possible under changeable conditions that we only associate with hunter-gatherer lifeways of boom, bust and flexibility (albeit the specifics will be warmer rather than colder)? Of course it is possible that sometime in the future a completely different subsistence base for humans will be developed – trophic shifts to dependance on insects, or algae, or seaweed, for example. But this chapter focuses on the near Anthropocene, asking what agriculture must or might look like. One of the most important means to re-frame agriculture is to break apart its monolithic status by demonstrating its spatial and temporal variability. Part of the reason it has hung together as a concept is because of what we have allowed it to say, and inferred from it, about the nature of being human.

To these ends, I will first revisit the debate that has occurred in archaeology about rethinking the phase of prehistory referred to as the 'Neolithic'. This illustrates the broader argument of the chapter by questioning understandings of agriculture as singular, stable and certain. And it helps us think further about our temporal divisions of the past. Are these phases of history – including the Anthropocene – understood as a convenient shorthand for capturing big picture, long-term change, or do they impose boundaries so strong that they delimit not only time but also our thinking? The critique of the Neolithic leads us into the broader conceptual critique in other disciplines of the concepts of agriculture and domestication. I also revisit the idea of 'bundles of practices', developed to deal with prehistoric change, and apply it to contemporary and future agriculture.

If variability, abundance and scarcity are major themes in the discussion of climate change, they have long been fundamental themes in agriculture. I draw on a variety of empirical evidence to consider what we can learn from farmer engagement with variability. It offers both resistance to, and a resource for, change. The final section of the chapter summarises the implications for both agriculture and the Anthropocene, particularly in helping both to be more open and flexible in relation to the future. It posits three experimental scenarios of Anthropocene agricultures.

Unpacking the agricultural assemblage

The Holocene is often presented as a benign and warm period, and that is true by comparison with the preceding Pleistocene. However at the local scale where people were – and still are – making decisions, it was much more variable and not always predictable. As Fagan (2004) argued, the stability of the Holocene climate has been overstated, at least at the scale at which human societies engage with it. Both small and large scale societies have struggled with and adapted to weather and climate throughout the Holocene; it has never been completely stable and certain.

Much recent Anthropocene debate is reminiscent of earlier discussions over the Neolithic Revolution. How did it come to be a dominant mode of life across much of the planet in the early to mid-Holocene? When and where did it start? What were the drivers, what were the responses and what are the reliable empirical indicators? In the last three decades such questions have been reframed in new approaches seeking to 'rethink the Neolithic' (Thomas 1991) – famously as neither Neolithic nor Revolution. In an ongoing careful exercise, scholars take issue with the questions, examine the empirical evidence more carefully and pay attention to the embedded concepts. I am not arguing that the Anthropocene and the Neolithic are similar, or even equivalent, phases of human history. I am arguing that they both represent pivotal changes in the way we understand and conceptualise human relations with the non-human world. Each discourse has its own distinctive politics, requiring us to consider whose voice counts. There are things to be learned, therefore, if the incipient intellectual community engaged in research into the Anthropocene reflects on how the transition to agriculture has been debated over recent decades (see Table 6.1 for a summary of dominant themes).

Gamble et al. (2005), for example, critiqued 'agricultural thinking' because of its inbuilt assumptions about origins, history and the processes of change. Our understandings of historical 'stages' and phases are themselves influenced by historical processes, and defined contingently in relation to one another. Depending how they are thought about, historical phases can replace one another, transition from one to another (Biermann 2014), be mutually embedded or just generally be messy. There are many different examples in current debates. For example, as Ruddiman (2013) argues, his early Anthropocene model has the additional awkward characteristic of swallowing most of the Holocene. And industrial capitalism is in large part an agricultural enterprise; like the anthropos, agriculture may be too big a category to have much explanatory traction.

A set of critical approaches have engaged with the implications of increasingly diverse empirical evidence for the concept of agriculture, and its companion concepts (hunter-gatherers, sedentism, civilisation, to name a few). Evidence increasingly showed that the various parts of the Neolithic 'package' did not all occur together, nor necessarily always in the same order. Sedentism sometimes preceded, sometimes followed agriculture.[3] Evidence has long shown agriculture to be a contingent emergence in a number of different ways (Davidson 1989,

Table 6.1 Key themes in rethinking the origins of agriculture

Theme	Examples
Critique of phase or stage-based approach to pre/history	Australian archaeological evidence challenged northern hemisphere stages.
	Arguments for contingent rather than progressivist models of archaeological change.
Many agricultural practices existed in so-called hunter-gatherer societies	Fruit seed germination on edge of campsites, yam cultivation, use of fire.
	Grinding stones occur both much earlier in time than agriculture, and persist in places where agriculture never appeared.
Empirical evidence of hunter-gatherer cultivation processes was often ignored or rendered invisible in the complex process of colonisation	This particularly applied to women's practices, such as tilling the soil to enhance the flourishing of tubers.
Archaeological evidence shows that agriculture emerged differently in different spaces and times	Evidence from yams, taros and bananas challenged the dominant Near Eastern 'cereal-centric' models, e.g. in Barton and Denham's (2011) 'vegecultures'.
	Independent emergence of agriculture in e.g. China, PNG, the Americas.
Conceptual critique of the concept of agriculture	From anthropology, with Ingold's (2000) articulation of 'dwelling', and from geography with Anderson's (1997) critique of animal domestication.

Source: Created using data from Head (2014) (where more detailed referencing can be found).

De Landa 1997). Contingent and nonlinear approaches (Terrell et al. 2003) – and their critique of grand syntheses and metanarratives – have come not only from rethinking the agricultural part of the equation, but also unravelling the monolithic concept of the 'precursor' hunter-gatherer phases. In the Australian context, for example, both Pleistocene and Holocene archaeological evidence suggest a 'past comprising a mosaic of independent cultural trajectories based on continuous adjustments' (Ulm 2013, p. 189) to local physical and social conditions (Hiscock 2008).

Perhaps the most important work in helping us understand how deeply the concept of agriculture is embedded in Western thought and practice is Kay Anderson's work, particularly her 2007 book *Race and the Crisis of Humanism*. As we discussed in Chapter 4, 'it was in the labour of cultivation that the cultural rather than natural character of the human was articulated' (Anderson 2007, p. 8). The colonisers' experience of Australia provided a fundamental challenge to this understanding, since:

> the non-farming nomad of Australia, who failed the lessons of improvement, stood apart according to the Enlightenment conception of humanness as an

essence and unity that realised itself in the movement out of nature ... The Australian was something 'other than human' where to be human was to be separate from nature.

(Anderson 2007, pp. 28–29)

In this reading, to be human is to lift oneself out of animality via the labour and process of cultivating the earth. Anderson went on to discuss the wider implications of this, for example the scripting of agriculture as the turning point that led to civilisation also led to the city conceptualised as an enclosing, or turning in from, nature.

The persistence of the agricultural trope is emphasised by the work of Gill (2014), who shows how it is articulated by pastoralists even in the arid rangelands of central Australia. Pastoralism is both different to agriculture, and subject to the same set of tropes:

Pastoralist claims to stewardship over time rely on representations of people, cattle and land that embody many of the tropes of improvement and domestication. This embodiment, however, is in often contradictory ways in which order is in tension with wildness as much as in transcendence.

(Gill 2014, p. 268)

In a series of key examples, Gill assembles pastoralist narratives of cattle as 'gardeners' who materially improve the country:

Pastoralists see the agency of cattle in soil disturbance that provides hoof indentations and breaks up the soil surface to facilitate seed and moisture penetration of the soil. Pastoralists referred to the role of cattle in 'opening up' vegetation and to seeing seeds and seedlings in hoof prints and vegetation growing on bare areas following the introduction of cattle to an area. This was evident in one older pastoralist's comment that stocking is 'just like ploughing the land' ...

In these narratives, cattle ... (re)create the landscape, they make it what it has become today. In their 'gardening', cattle mimic and enhance natural processes. Cattle enact stewardship; they are not only good for country, they make country good. In so doing, they echo more linear pathways of improvement through cultivation for agriculture and rein in agrios in favour of a pastoral domus, but in a form distinct from that embodied in living with chaos.

(2014, p. 271)

I return below to the way Gill's pastoralists engaged with rainfall variability and its relationship to chaos.

Bundles of 'planty' practices

Chase famously called Australian Aboriginal practices 'domiculture', the 'knowledge and activity bundles which relate temporally to a specific habitat' (1989, p. 48). Building on this, Denham et al. (2009) conceptualised human-plant relations over archaeological timescales as constituted by 'bundles of practices', reminding us that close empirical attention to variation in space and time reveals very different patterns to the imposition of pre-constituted categories. We suggested that:

> the key to understanding the emergence of variation is conceptualising how common sets of technologies and practices were bundled in given locales and then how these transformed through time, either through innovation or adoption. Bundling refers to how different practices co-occur in social, spatial and historical contexts; it invokes a contingent collection without permanence or necessary relationship.
>
> (2009, p. 35)

Thus, for example, practices such as landscape burning, tree exploitation, tuber exploitation, digging, transplanting and detoxification were bundled – contemporary cultural geographers might say assembled – differently in different parts of Australia and New Guinea, leading to the emergence of regionally distinctive collectives by the late Holocene.

Building the empirical evidence of those practices and their variability is an ongoing multidisciplinary enterprise, and many of the details change. Increasingly, however, evidence shows that particular practices have been present for long periods of time before they become part of an agricultural package. The grinding of seeds and ground-stone implements (milling stones and axes) – the supposedly defining components of the 'Neolithic' – provide perhaps the most dramatic example. Use-wear and residue analysis of stone tools from the Madjedbebe site in northern Australia shows clearly that the technology for ground-edge axes, and other grinding practices, were part of the bundle that Australia's first colonisers brought with them (Clarkson et al. 2015). Evidence shows that seed-grinding practices were present at least 20,000 years ago in Europe, Asia and Australia (Fullagar et al. 2015), and considerably earlier in Africa (Mercader 2009). These arguments for variability apply not only to the variable onset of agriculture but also to its later manifestations.

If we are to think of human-plant relations over agricultural timescales as constituted by 'bundles of practices', it follows that we must do so for the hunter-gatherer package as well. Palaeoecological, anthropological, historical and contemporary ethnographic evidence is replete with examples of diverse practices, including the important Australian example of landscape burning (Jones 1969, Gammage 2011, Mooney et al. 2011). Many of these practices have been ignored by different observers over time because they did not fit the categorisations with which people were seeing the world. Invisibility of planting and soil practices was partly to do with their gendered nature; the descriptions are overwhelmingly of women's work (Gott 1982, 1983). Indigenous Australian communities have partly suffered from attempts to generalise their societies and economies at the continental scale, whereas nuanced and careful regional studies (Keen 2004) show considerable variability.

Even within a particular food category, there are diverse implications for landscape transformation and social engagement. One example of note comes from yams, a loose category used to refer to plants with underground storage organs, often implicated in discussions over the extent to which Aboriginal people were 'gardeners'. In our overview of the biogeography and social life of Australian *Dioscorea* yams, Jenny Atchison and I (Atchison & Head 2012) showed that the geographic variability of yam species across the continent has important consequences for the scale and intensity of Aboriginal collection and landscape transformation. On the alluvial terraces of the coastal plains of southwestern Australia, extensive *D. hastifolia* (*warran*) grounds were recorded by early European explorers:

> for three and a half consecutive miles we traversed a fertile piece of land, literally perforated with the holes the natives had made to dig this root; indeed we could with difficulty walk across it on that account, whilst this tract extended east and west as far as we could see.
> (Grey 1841, p. 12, in Hallam 1986, p. 118)

Grey commented that 'superior huts, well marked roads, deeply sunk wells, and extensive warran grounds, all spoke of a large and comparatively speaking resident population' (Grey 1841, p. 20).

In contrast, *D. transversa* and *D. bulbifera* grounds in northwestern Australia occur within monsoonal vine thicket patches on rocky slopes. The process of 'gardening' in such habitats requires a kind of quarrying; the removal of tonnes of rock and soil during the tuber excavation process, which also interacts with the process of quarrying for stone artefact sources (Head et al. 2002). These impacts are localised and patchy, affecting areas as small as tens of square metres.

In a different kind of spatially fine-grained study, Atchison (2009) examined stand structure and recruitment of the fruit trees *Persoonia falcata* and *Buchanania obovata* around four sites in northwestern Australia, documented as important Aboriginal traditional and post-contact wet-season camping places. She showed that seedlings of *P. falcata* were only recruited to adult life stages in the one site where Aboriginal people had re-introduced customary management practices via burning.

It is true that many such practices are not visible at common scales of landscape analysis. Anthromes, for example, have a spatial resolution on the ground of about 85 km^2, so they will not pick up micro-practices unless they visibly change the land cover at broader scales. We can only wish to capture the spectral signatures reflected from the Earth 8000 or 40,000 years ago! However, at the microscopic scale, genetic evidence is throwing new light on patterns of plant distribution previously assumed to be 'natural' or 'relict', as in not affected by the presence of people.

Rangan et al. (2015) combined genetic and historical linguistic data to argue that humans were the most likely vectors of boab (*Adansonia gregorii*) movement across long distances and biogeographic barriers (between the Kimberley to the west and Arnhem Land to the east, and within the Kimberley as constituted by major river drainage systems) in northwestern Australia. Boabs had a range of uses recorded ethnographically, and an archaeologically demonstrated history of use dating back to 39,000 years ago, with substantial increases from 3000BP onwards, peaking 650 years ago (McConnell & O'Connor 1997, 1999), but were not domesticated or cultivated in any conventional sense. 'Given the lack of other obvious seed dispersal agents, the lack of evidence of barriers to gene flow is most easily explained by a long history of humans moving boab seeds' (Rangan et al. 2015, p. 10).

In an example of *Livistona* palms in central Australia, long assumed in Australian biogeography to be relicts of Cenozoic aridification, Kondo et al. (2012) used genetic methods to rule out the relict hypothesis and suggest that human dispersal vectors were as likely as animal ones. They concluded that:

> It is likely that the *L. mariae* populations were established by opportunistic immigrants via long-distance seed dispersal over the 1000 km gap . . . it is as likely that it is a legacy of Aboriginal dispersal as it is that it was carried there by animals.
>
> (2012, p. 2659)

For me the most interesting thing about the new genetic clues in the cases of boabs and *Livistona* palms is not the specifics of the scientific findings, which may be modified by future research. Nor is it the diversification of the way we picture categories such as gardens, agriculture or gathering. That diversity has been long-standing in the historical and ethnographic literature, if only we were prepared to look. Rather the thing that is notable in these and recent studies (or indeed in the absence of specific studies) is the persistence of the default explanation that humans, specifically hunter-gatherer humans, would not be an agent of change in this way. This is considered the thing that scientists have to disprove; Kondo et al. (2012) have to go to some length to explain why a human agent of dispersal is as likely as birds or bats. Here we are reminded of how deeply both ecological and agricultural explanations differentiate humans, the former by their absence, and the latter by their presence.

So, in focusing on human practices and transformations of the landscape in this section, I perhaps risk too much of a human-centred view of the world. It is certainly true that we have much to gain from considering the agricultural endeavour as a co-production of humans, animals, plants and many other contributors. But it is also important to pause and reflect on the extent to which particular framings of the human place in nature, and particular framings of particular groups of humans, and particular framings of what it means to be in/out of nature, have constrained and limited our agricultural thinking. We cannot go on to develop a perspective that draws in a more inclusive depiction of co-production until we have absorbed and understood this.

Variability – source of both resistance and skill

In exploring variability and its relationship to agriculture, we should also understand it as a diverse package comprising many human and other elements. Part of my provocation comes from Hulme (2015), who discusses climate as a concept stabilising and normalising weather – the volatile everyday experiences of variable wind, rain and sun. Climate:

> offers a way of navigating between the human experience of a constantly changing atmosphere and its attendant insecurities, and the need to live with a sense of stability and regularity ... The idea of climate cultivates the possibility of a stable psychological life and of meaningful human action in the world. Put simply, climate allows humans to live culturally with their weather.
>
> (Hulme 2015, p. 3)

In the context of climate change, 'climate will be called upon to do new work in new circumstances, to meet the enduring human need for order, security and meaning' (Hulme 2015, p. 5). However, Hulme also emphasises, such stability can only ever be aspirational, since climates can never be either physically or imaginatively stable. 'The idea of climate brings order to human life only in so far that it continually accommodates change in its constituent elements of weather and culture' (Hulme 2015, p. 6). We will need to renegotiate these relationships, but with long-standing cultural resources: 'climate change should not be understood as a decisive break from the past nor as a unique outcome of modernity' (Hulme 2015, p. 10).[4]

One avenue of hopeful futures is to contest any presumption that climate stability is a necessary state for civilisation, and instability a bad thing in itself. Hulme himself reviews diverse evidence showing how drought and aridification can be a potential trigger for increased social complexity:

> This reading of the evidence also sits with more nuanced views that argue that the outcome of any given climatic shift is deeply dependent on the resilience and adaptability of local societies and institutions. There is no simple mapping between climate change and the fortunes of civilisations.
>
> (2009, p. 31)

We can extend this thinking about variability to agriculture itself. Agriculture can be understood as a stabilising concept and practice; it attempts to smooth out variability, concentrates productivity seasonally and facilitates storage. Irrigated agriculture is even more so. It has to capture, store and redistribute water in order to synchronise it with the needs of the plants being grown. Further, this variability is not all climatic, but incorporates unevenness or friction in transport systems, logistics and markets. So, as Taylor emphasises, the influences of weather, climate and climate change on agriculture are not unmediated:

a change in material climate can be produced through re-ordering the socio-ecological relations at regionalised or local levels that affects the way that meteorological forces are inhered within the landscape. Consider, for example, an agrarian community undergoing a shift driven by colonial duress from pastoral practices on common rain-fed lands to a form of settled agriculture with privatised property and canalised irrigation ... Through such socio-ecological ruptures, both the material expressions and lived experiences of climatic processes are profoundly transformed.

(2015, pp. 41–42)

Thus, for example, wheat farmers in Australia and Russia each often do well when the other has a poor season due to drought. Horticulturalists using irrigation in Australia's Sunraysia region may experience reduced runoff upstream through pricing mechanisms rather than an actual shortage of water. Boom and bust ecologies or societies are about abundance as well as scarcity. Water is one example. In a crude sense, we are going to have more in northern Australia and less in southern Australia, where most of the people are, under climate change scenarios.

On the face of things, engagement with variability is precisely what Australians do well, as illustrated vividly in the way we talk about climatic variability. We are very quick to invoke Dorothea Mackellar's 'droughts and flooding rains' to illustrate our acceptance of change and extremes as normal, for example in relation to the bushfires discussed in Chapter 2. Research shows such stoicism to be a recurrent trope among farmers. One of Gill's interviewees, for example, 'emphasised the power of natural processes in driving change – "you can't hold back nature" – and the long timescales of observation needed to have any chance of comprehending what was occurring' (2014, p. 270).

So, normalisation of variability can be a tool of stabilising change, and thus resisting climate change science. It is also a highly gendered thing; affirming the stoic, resilient, tough Australian (or any other) male mode of coping. But can it also provide a cultural resource for dealing with precisely the kind of increased variability that we might expect under climate change scenarios? As Gill has noted, the narratives he recorded among arid zone pastoralists 'could readily be used to valorise local knowledge, long-term commitment to place, humility in the face of nature and a commitment to coexist with unpredictability' (2014, p. 270).

It is notable that similar narratives have been recorded across very different farming types in Australia. Gill's pastoralists live and work in the Australian arid zone, one of the defining features of which is rainfall variability. Wheat farmers in New South Wales expressed a range of views about 'normal' variability, and whether recent drought is part of this range, or 'something's going on' (Head et al. 2011, p. 1102). Dryland farmers in South Australia invoked the normality of climate variability to contest the science of climate change (Raymond & Robinson 2013). In both the latter studies, farming households are in a continuous process of managing risk, not just weather and climate risks, but also those related to finances, investment in technology and household wellbeing. Longer-term adaptation is mediated through short-term decision-making, and in some ways it cannot be otherwise. Wheat farmers, for example, have to make decisions about which varieties to plant at the beginning of the season. Regardless of seasonal or longer-term predictions, each year's crop and harvest is the outcome of a set of decisions locked in before planting.

This view of variability is also echoed by irrigated grape growers in the Sunraysia district of Australia's Murray valley, around Mildura and Robinvale.[5] In an interview discussion of climate change, wine grape growers Brian and Rosalie said, 'I think climate is going to continue to change. It has over the many centuries' (Brian) and 'It's a cyclic thing' (Rosalie). They then turned the discussion to the ongoing work to develop new plant varieties to cope with changing conditions. Dried-fruit grower Keith said:

> Look, I think climate change is probably the case, to some extent, and I suppose you've got to balance that out about what would occur anyway and whether man's done it or not. With the amount of fossil fuel taken out of the earth I suppose you've got to assume that's going to have some effect on things. I'm not sure it'll make things hellish and different here, but how would I know? But the climate here has been so volatile, I mean it's so erratic. I mean I haven't seen any evidence that things have changed. I mean people say, 'Oh, we had the big floods, and we've had the big droughts, we've had this, that and the other.' We've always had them, and whether you can say it's attributable to climate change, I don't know ... you know, our weather is just so erratic it's just hard to pinpoint any specific change. So I guess we'll continue to battle with erratic weather, whether it's climate change or not.
>
> (Keith, Mildura, May 2015)

The climatic variability under discussion here is not all about storing water for scarcity, it is also about worrying that it might fall at the wrong time. Paradoxically, in this semi-arid region growing water-intensive crops, it is possible to also talk of the 'risk of rain'. Grape farmers throughout the area put plastic covers as raincoats over vines to prevent potential damage (Figure 6.2). Keith's discussion of this illustrates one of the ways his negotiation with climate change is mediated through short-term weather. For dried fruit growers, the drying of the grapes in the lead up to and immediately after harvest is a crucial variable in the quality of the crop and thus the price it will fetch. In the course of the interview Keith discussed historical changes towards mechanisation in the cutting and drying processes:

> It's still influenced by the rain. So what we do nowadays is, and I don't think we can ever get over the risk of rain; last year we ran some trials through the Dried Fruits Association on putting covers and things like that on, like the table grape growers do, but in our instance we can produce some really

Figure 6.2 White plastic protects grape vines near Euston, NSW. Photo: ARC DP140101165.

top quality fruit that might get rained on but it'll usually go brown, and it's only a colour issue. So the German market likes good, light coloured fruit . . .

. . . we try to minimise the time it's out there hanging in the elements so that the risk of it getting rained on is less by the sheer fact that it's out there for less time. And so if we can dry it within, say, 15, 20 days that's about as good as you can do.

(Keith, Mildura, May 2015)

The risks of rain are slightly different for dried, table and wine grapes, but include splitting and discolouration of the fruit, and mould and fungal growth due to increased humidity.

Keith and his wife Jenny, Brian and Rosalie are all approaching retirement. None of their children want to continue the farms, and issues of farm futures loom large, in both spoken and unspoken ways. Participants in Raymond and Robinson's (2013) study expressed this generalised fear of the future around family farming in Australia. Boosterish government projections contrast with the reality of an ageing farm workforce, and limited opportunities for those with non-farming backgrounds to be able to buy into land. Anxiety about climate change is only part of this wider anxiety:

I think the collective view would be confusion. Everyone knows that climate variability is here; we live in a variable climate, but I think they would individually acknowledge that climate change has or is or will occur, but they are very easy to talk out of it. They would be hanging on anyone who says it's not going to happen, because they're scared of the future. There's definitely absolute fear of the future. One of those ways to handle that fear is ignore it, debunk it, just call it lies, but given that profitability is marginal at current prices and production, everyone is scared of the future.

(Institutional respondent, Raymond & Robinson 2013, p. 107)

Whether it is true, as Hulme (2015) implies, that humans need certainty, or can learn to live with increased levels of uncertainty, the goal is pretty similar. Can we destabilise or expand our view of normality sufficiently to deal with the challenges ahead of us? Can we articulate resilience in a way that is transformative rather than conservative? We know there are problems with the Dorothea Mackellar view of the world. For one thing it has the 'droughts and flooding rains' out there, as separate processes that are 'done to' people. Second, it reflects an essentially linear view, in which normal is understood as something between two extremes. Just stretching out the bell curve to encompass a broader field of normal only takes us so far. We need to incorporate non-linear change into this understanding, and that is a kind of change that applies not only to climate models but also to social systems. A major issue needing consideration is applying concepts of variability that resonate outside the farming context. One of the climate scientists interviewed for the project discussed in Chapter 5 acknowledged this double-edged sword of the way farmers deal with variability, and then went on to note that the real issue is with urban communities: 'I don't think most people in urban areas think we live in a highly variable environment really. I don't think it is a very deep narrative in the wider, and in fact the majority of Australian society' (Paul).

Rethinking agriculture for the Anthropocene, and rethinking the Anthropocene

The examples in this chapter are diverse, perhaps too diverse. On one hand the agricultural package is so variable that it can encompass both Aboriginal yamming practices and petro-industrial wheat production. On the other, I have argued that the subsistence underpinnings of the Anthropocene will be constituted by practices that we recognise as agricultural. So at the risk of contradicting myself, let us consider what the characteristics of Anthropocene agriculture might be. (We can leave aside for the moment whether it is best to call it agriculture, or whether a series of other names might be better.) Or, to put it more positively, we can consider how reconceptualising agriculture opens up new possibilities. Here I speculate on three potential, or experimental, bundles of practices.

The bundle of ethnic diversity and land tenure practices

There are new farmers in the Sunraysia region, such as the corporate almond producers who employ seasonal labourers from Pacific communities. There are viable successions, such as families of Italian post-WWII migrants who continue to be involved in vegetable and grape growing. It is no accident that many refugees and new migrants are directed to this rich agricultural area – it has an ongoing need for labour. This concentration of migrants makes the Sunraysia region one of the most ethnically diverse parts of rural Australia; one third of horticulturalists in the broader region speak a language other than English at home. Many bring considerable agricultural knowledge with them, and should be considered part of Australia's farming future.

Consider Baba, a Hazara refugee, born in Afghanistan. He was a farmer, until his farm was taken by the Taliban, and he had to flee to Pakistan. Baba described a mixture of subsistence crops such as wheat, fruit and vegetables, and market crops, the main one being almonds. At the time of interview, he had been in Australia nearly four years, most of them in Mildura and Swan Hill. Asked, through an interpreter, for his first impressions of Australia,[6] he talked about Australia as a place of freedom, where the rule of law applies in practice, not just on paper. Asked then about his first impressions of the Australian environment, Baba commented on the way water is used in agriculture, particularly via drip irrigation. In Afghanistan and Pakistan, by contrast, where he was dependent on snow melt for spring water supplies, people were more likely to 'waste' water: 'From a water pond we water fifty trees, my brother. Here they water five hundred trees. They use it drop by drop'. 'Interesting you noticed about the water here', responded the interviewer. Baba drew on his status and knowledge as a farmer to reply, comparing himself to a painter who looks at the paint, or an engineer who looks at the building.

His friend Nader, also a farmer in Afghanistan, who had been in Australia for a little over a month at the time of our interview, described the process by which his village diverted water from the snow melt stream into a pool, then decided how to take turns to channel the water onto their crops. The context they describe from home in Afghanistan is of water management that harnesses seasonal water variability – the provision of snow melt – through a combination of technologies such as dug canals and storage ponds, and social practices of sharing and conflict mediation. The specifics are different, but the broad outline is in many ways similar to the dominant narrative of irrigated agriculture in the Mildura-Robinvale area.

The chances of Baba and Nader becoming farmers in Australia – assuming they even wanted to – seem negligible, due to the high capital costs of land and water. It is unlikely that land will be stolen from its previous owners to give to them, as it was stolen from Indigenous owners to facilitate nineteenth-century pastoralism and agriculture. For most of the migrants we interviewed in this area, the possibility of becoming a farmer was so remote they did not even dare to dream of it. The main dream of those who are refugees is just that their family will be able to join them in Australia. There are of course major questions over the long-term viability of both the Himalayan snow melt and irrigation in the Murray-Darling system. To suggest that climate change is an impending catastrophe in both areas is perhaps to belittle the absolute horrors that people such as Baba and Nader have left behind in their home countries. It is definitely to talk from a position of privilege and affluence. So it may seem premature or inappropriate for us to ponder alternative models of agricultural futures in the Murray-Darling system. But as a research group we have been struck by the abandonment of some agricultural land and the fact that diverse groups with farming skills desperately need agricultural land. Are there possibilities for the holes in this 'swiss cheese landscape', as one grape grower described it, being filled in with diverse farming practices that look completely inefficient within the current model of productivity? These could include subsistence as well as market practices. As Taylor has argued for the Pakistani context, 'land reform might be considered a "climate change adaption" policy par excellence' (2015, p. 138).

In the examples provided above, people are responding to, and are part of, complex assemblages of socio-eco-techno-water. Neither Baba nor Nader perceive or respond to Australia as the world's driest inhabited continent. In fact they both express the opposite – an encounter with a paradise of unlimited and well-managed supply. To the extent that water frugality is discussed in these interviews, it is something brought from previous experience that may be gradually let go, rather than a new response to living on an arid continent. As migration flows increase across the globe, many more people will encounter and find themselves living in new socio-ecological assemblages. This provides both opportunity to do things differently, and challenges to do them well.

The landscape bundle

The kinds of examples that can be distilled from migrant experiences make important contributions to unsettling and destabilising our monolithic view of agriculture. One important reason is that in all our Robinvale-Mildura interviews, the lens on the environment is through the production of food. That is, when people are asked about environmental issues, they answer through the lens of food production, and particularly the importance of water. We must still recognise that in the Australian context, agriculture is held somewhat separate from what is generally understood to be environmental thought and practice. It is one of the luxuries of our affluence (as well as the colonial inheritance) that we have held them separate in quite artificial ways. In places such as Sweden, where agricultural history is recognised as constituting a longer part of the cultural history, a different sort of welcome is accorded agricultural plants, animals and landscapes into the environmental fold (Saltzman et al. 2011). We have discussed previously what it might mean to allow Australian crops such as wheat to belong in Australian nature (Head et al. 2012). This is not to deny the many environmental problems associated with agricultural production, but we will not solve them by continuing to conceptualise 'environment' and 'agriculture' as two separate domains. Across

Rethinking agriculture, rethinking Anthropocene 109

Figure 6.3 Rice farms at the edge of Tuckerbil Swamp, 2012. Ramsar listed wetland, NSW. Photo: Emily O'Gorman.

the landscape scale, these separationist versus associationist approaches to conservation are sometimes argued to be the difference between land sparing and land sharing (Tscharntke et al. 2012). A further take on this is Emily O'Gorman's (2014) work on the relationships between native ducks and rice fields. O'Gorman challenges the concept of wetland, understood conventionally as a native, pre-European landscape. In showing the spatial, practical and sometimes conflictual relationships, she extends the concept of wetlands to include rice agriculture (Figure 6.3).

In these understandings, a much wider set of plants and animals must be welcomed, must be considered to belong, as part of the agricultural package, and/ or to belong in the Australian landscape. The challenges to categories of belonging are discussed further in Chapter 7.

The Typha *bundle*

Typha is not a superfood. It takes a lot of work to dig it. The rhizome tastes best in late autumn and early winter when richest in starch. Interpreting the full range of ethnographic sources, Gott argued that 'Typha was available year-round if necessary, but was more nourishing and attractive at certain seasons, implying, of course, that the food supply was sufficiently ample to allow choice to be made' (1999, p. 41).

The bundle of practices that traditionally accompanied *Typha* cultivation included burning in late winter, harvesting, pounding, cooking, eating, chewing, rolling string on the thigh, making nets. The Anthropocene equivalents could include all of these and more; damming polluted waterways to encourage its cleansing presence, or creating unpolluted dams – or backyard ponds – to farm *Typha* patches. It could provide emergency food sources.

The *Typha* bundle is literally a bundle if we think of the usefulness of its fibres. Gott reports the string preparation process as follows:

> After the starch was chewed off, the remaining fibre was packed in bags which were temporarily used as pillows. When required for string-making, the knots were soaked overnight in water, teased out and scraped with shells of the freshwater mussel until they were clean, after which they were stored in small neat hanks ready to be made into two-ply string.
>
> (1999, p. 41)

Insights from Indigenous practices can be applied to the processes and practices of storage and transport, as well as to growing. Here we may draw on the idea of storage in the landscape – people did not have to pick yams or *Typha* all the time because they knew precisely where they were under the ground. If we reinstitute subsistence gardening in the suburbs, we can think of storage in the landscape. Once again, attention to vernacular practices and the fine-grained detail of everyday life shows that the seeds of possibility are already present.

These three experimental bundles of practices seem far-fetched when viewed through the conventional frame of modernist agriculture. Yet that frame contains its own extremes and contradictions, for example that there is a food security issue in the Sunraysia region,[7] one of the food bowls of the Murray-Darling system. Or that irrigated agricultural production in the Murray-Darling might decline by 92 per cent without climate change mitigation (Garnaut 2008). *Typha* practices challenge several divides; between native plants and agriculture, environment and agriculture and urban and rural. What they also provide for us is one way to draw on the long heritage of this continent and Indigenous knowledge as a resource for common futures. It carries into the Anthropocene things some of us already know how to do. All of this must surely show us that, however we think of it, the Anthropocene is not something that comes after, not a stage or layer of history. It is something in which we are already entwined.

Notes

1. Australia has two native species of the cosmopolitan genus *Typha*: *T. domingensis* (narrow-leaved cumbungi), more common in inland areas, and *T. orientalis* (broad-leaved cumbungi), more common in coastal and tableland areas. The northern hemisphere species *T. latifolia* has become naturalised in some areas since European settlement (Gott 1999).
2. A variety of other uses is reported by Gott (1999) and Clarke (2012). These include stems being used to make portable fish traps and flower stems for spear shafts. Down from flowers was used to line bark wallets, decorate dancers during ceremonies and stuff pillows.

3 I am not concerned here with the specifics of the (considerable) methodological or empirical disagreements in those debates, except insofar as they throw broader interpretive questions into relief. For examples of the broad range of views on both concept and method, see Bellwood et al. (2007).
4 Note in this discussion I am avoiding the concept of resilience. I have written previously (Head 2012) about my concern that the concept of resilience emphasises conservatism and stability rather than transformative change. Others have critiqued the concept much more comprehensively (Walker & Cooper 2011, Evans & Reid 2014, Watts 2015).
5 The Mildura and Robinvale research is being undertaken in collaboration with Natascha Klocker and Olivia Dun. It is part of a broader project on ethnic diversity and sustainability/climate change issues, together with Gordon Waitt and Heather Goodall (Klocker & Head 2013).
6 Interview by Olivia Dun.
7 www.healthytogether.vic.gov.au/blog/posts/creating-a-healthier-food-system-in-mildura. Last accessed 25 August 2015.

References

Anderson, K. 1997. "A walk on the wild side: A critical geography of domestication." *Progress in Human Geography* 21(4): 463–485.

Anderson, K. 2007. *Race and the Crisis of Humanism*. London: UCL Press.

Atchison, J. 2009. "Human impacts on *Persoonia falcata*. Perspectives on post-contact vegetation change in the Keep River region, Australia, from contemporary vegetation surveys." *Vegetation History and Archaeobotany* 18(2): 147–157.

Atchison, J. and L. Head. 2012. "Yam landscapes: The biogeography and social life of Australian *Dioscorea*." *The Artefact* 35: 59–74.

Barton, H. and T. Denham. 2011. "Prehistoric vegeculture and social life in Island Southeast Asia and Melanesia." In *Why Cultivate? Anthropological and Archaeological Approaches to Foraging-Farming Transitions in Southeast Asia*, edited by G. Barker and M. Janowski, 17–25. Cambridge: University of Cambridge.

Bellwood P., C. Gamble, S. A. Le Blanc, M. Pluciennik, M. Richards and J. E. Terrell. 2007. "Review feature: *First farmers: The origins of agricultural societies, by Peter Bellwood*." *Cambridge Archaeological Journal* 17(1): 87–109.

Biermann, F. 2014. "The Anthropocene: A governance perspective." *The Anthropocene Review* 1: 57–61.

Chase, A. K. 1989. "Domestication and domiculture in northern Australia: A social perspective." In *Foraging and Farming: The Evolution of Plant Exploitation*, edited by D. R. Harris and G. C. Hillman, 42–54. London: Unwin Hyman.

Clarke, P. A. 2012. *Australian Plants as Aboriginal Tools*. Dural, NSW: Rosenberg Publishing.

Clarkson, C., M. Smith, B. Marwick, R. Fullagar, L. A. Wallis, P. Faulkner, T. Manne, E. Hayes, R. G. Roberts, Z. Jacobs, X. Carah, K. M. Lowe, J. Matthews and S. A. Florin. 2015. "The archaeology, chronology and stratigraphy of Madjedbebe (Malakunanja II): A site in northern Australia with early occupation." *Journal of Human Evolution* 83: 46–64.

Davidson, I. 1989. "Is intensification a condition of the fisher-hunter-gatherer way of life?" *Archaeology in Oceania* 24(2): 75–78.

De Landa, M. 1997. *A Thousand Years of Nonlinear History*. New York: Zone Books.

Denham, T., R. Fullagar and L. Head. 2009. "Plant exploitation on Sahul: From colonisation to the emergence of regional specialisation during the Holocene." *Quaternary International* 202: 29–40.

Ellis, E. C. 2013. "Sustaining biodiversity and people in the world's anthropogenic biomes." *Current Opinion in Environmental Sustainability* 5: 368–372.

Ellis, E. C. and N. Ramankutty. 2008. "Putting people in the map: Anthropogenic biomes of the world." *Frontiers in Ecology and the Environment* 6(8): 439–447.

Evans, B. and J. Reid. 2014. *Resilient Life: The Art of Living Dangerously*. Cambridge: Polity Press.

Fagan, B. 2004. *The Long Summer: How Climate Changed Civilization*. London: Granta Books.

Fullagar, R., E. Hayes, B. Stephenson, J. Field, C. Matheson, N. Stern, K. Fitzsimmons. 2015. "Evidence for Pleistocene seed-grinding at Lake Mungo, south-eastern Australia." *Archaeology in Oceania* 50: 3–18.

Gamble, C., W. Davies, P. Pettitt, L. Hazelwood and M. Richards. 2005. "The archaeological and genetic foundations of the European population during the Late Glacial: Implications for 'agricultural thinking'." *Cambridge Archaeological Journal* 15(2): 193–223.

Gammage, B. 2011. *The Biggest Estate on Earth: How Aborigines Made Australia*. Sydney: Allen & Unwin.

Garnaut, R. 2008. *The Garnaut Climate Change Review*. Cambridge: Cambridge University Press.

Gill, N. 2014. "Making country good: Stewardship and environmental change in central Australian pastoral culture." *Transactions of the Institute of British Geographers* 39(2): 265–277.

Gott, B. 1982. "Ecology of root use by the Aborigines of southern Australia." *Archaeology in Oceania* 17(1): 59–67.

Gott, B. 1983. "Murnong – *Microseris scapigera*: A study of a staple food of Victorian Aborigines." *Australian Aboriginal Studies* 2: 2–18.

Gott, B. 1999. "Cumbungi, *Typha* species: A staple Aboriginal food in southern Australia." *Australian Aboriginal Studies* 1: 33–50.

Grey, G. 1841. *Journals of Two Expeditions of Discovery in North-West and Western Australia, During the Years 1837, 1838 and 1839*. London: T and W Boone.

Hallam, S. J. 1986. "Yams, alluvium and villages on the west coastal plain." At the *54th Congress of the Australian and New Zealand Association for the Advancement of Science (ANZAAS)*. May 1986. Australian Institute of Aboriginal Studies, Canberra.

Head, L. 2012. "Conceptualising the human in cultural landscapes and resilience thinking." In *Resilience and the Cultural Landscape: Understanding and Managing Change in Human-Shaped Environments*, edited by T. Plieninger and C. Bieling, 65–79. Cambridge: Cambridge University Press.

Head, L. 2014. "Contingencies of the Anthropocene: Lessons from the 'Neolithic'." *The Anthropocene Review* 1(2): 113–125.

Head, L., J. Atchison and R. Fullagar. 2002. "Country and garden: Ethnobotany, archaeobotany and Aboriginal landscapes near the Keep River, northwestern Australia." *Journal of Social Archaeology* 2(2): 173–196.

Head, L., J. Atchison and A. Gates. 2012. *Ingrained: A Human Bio-Geography of Wheat*. Aldershot: Ashgate.

Head, L., J. Atchison, A. Gates and P. Muir. 2011. "A fine-grained study of the experience of drought, risk and climate change among Australian wheat farming households." *Annals of the Association of American Geographers* 101(5): 1089–1108.

Hiscock, P. 2008. *Archaeology of Ancient Australia*. Abingdon: Routledge.

Hulme, M. 2009. *Why We Disagree About Climate Change: Understanding Controversy, Inaction and Opportunity*. Cambridge: Cambridge University Press.

Hulme, M. 2015. "Climate and its changes: A cultural appraisal." *Geography and Environment*, DOI: 10.1002/geo2.5.

Ingold, T. 2000. *The Perception of the Environment: Essays on Livelihood, Dwelling and Skill*. Abingdon: Routledge.

Jones, R. 1969. "Fire-stick farming." *Australian Natural History* 16(7): 224–228.

Keen, I. 2004. *Aboriginal Economy and Society: Australia at the Threshold of Colonisation*. South Melbourne: Oxford University Press.

Klocker, N. and L. Head. 2013. "Diversifying ethnicity in Australia's population and environment debates." *Australian Geographer* 44(1): 41–62.

Kondo T., M. D. Crisp, C. Linde, D. M. J. S. Bowman, K. Kawamura, S. Kaneko and Y. Isagi. 2012. "Not an ancient relic: The endemic *Livistona* palms of arid central Australia could have been introduced by humans." *Proceedings of the Royal Society B*. 279: 2652–2661.

McConnell, K. and S. O'Connor. 1997. "40,000 year record of food plants in the Southern Kimberley Ranges, Western Australia." *Australian Archaeology* 45: 20–31.

McConnell, K. and S. O'Connor. 1999. "Carpenter's Gap Shelter I: A case for total recovery." In *Taphonomy: The Analysis of Processes from Phytoliths to Megafauna*, edited by M.-J. Mountain and D. Bowdery, 23–34. Canberra: ANH Publications.

Mercader, J. 2009. "Mozambican grass consumption during the Middle Stone Age." *Science* 326: 1680–1683.

Mooney, S. D., S. P. Harrison, P. J. Bartlein, A.-L. Daniau, J. Stevenson, K. C. Brownlie, S. Buckman, M. Cupper, J. Luly, M. Black, E. Colhoun, D. D'Costa, J. Dodson, S. Haberle, G. S. Hope, P. Kershaw, C. Kenyon, M. McKenzie and N. Williams. 2011. "Late quaternary fire regimes of Australasia." *Quaternary Science Reviews* 30: 28–46.

O'Gorman, E. 2014. "Remaking wetlands: Rice fields and ducks in the Murrumbidgee River Region, NSW." In *Rethinking Invasion Ecologies From the Environmental Humanities*, edited by J. Frawley and I. McCalman, 215–238. Abingdon: Routledge.

Rangan, H., K. L. Bell, D. A. Baum, R. Fowler, P. McConvell, T. Saunders, S. Spronck, C. A. Kull and D. J. Murphy. 2015. "New genetic and linguistic analyses show ancient human influence on Baobab evolution and distribution in Australia." *PLOS One*. DOI: 10.1371/journal.pone.0119758.

Raymond, C. M. and G. M. Robinson. 2013. "Factors affecting rural landholders' adaptation to climate change: Insights from formal institutions and communities of practice." *Global Environmental Change* 23(1): 103–114.

Ruddiman, W. 2003. "The atmospheric greenhouse era began thousands of years ago." *Climate Change* 61(3): 261–293.

Ruddiman, W. 2013. "The Anthropocene." *The Annual Review of Earth and Planetary Sciences* 41:45–68.

Saltzman, K., L. Head and M. Stenseke. 2011. "Do cows belong in nature? The cultural basis of agriculture in Sweden and Australia." *Journal of Rural Studies* 27(1): 54–62.

Taylor, M. 2015. *The Political Ecology of Climate Change Adaptation: Livelihoods, Agrarian Change and the Conflicts of Development*. Abingdon: Routledge.

Terrell, J. E., J. P. Hart, S. Barut, N. Cellinese, A. Curet, T. Denham, C. M. Kusimba, K. Latinis, R. Oka, J. Palka, M. E. D. Pohl, K. O. Pope, P. R. Williams, H. Haines and J. E. Staller. 2003. "Domesticated landscapes: The subsistence ecology of plant and animal domestication." *Journal of Archaeological Method and Theory* 10(4): 323–368.

Thomas, J. 1991. *Rethinking the Neolithic*. Cambridge: Cambridge University Press.

Tscharntke, T., Y. Clough, T. C. Wanger, L. Jackson, I. Motzke, I. Perfecto, J. Vandermeer and A. Whitbread. 2012. "Global food security, biodiversity conservation and the future of agricultural intensification." *Biological Conservation* 151(1): 53–59.

Ulm, S. 2013. "'Complexity' and the Australian continental narrative: Themes in the archaeology of Holocene Australia." *Quaternary International* 285: 182–192.

Walker, J. and M. Cooper. 2011. "Genealogies of resilience: From systems ecology to the political economy of crisis adaptation." *Security Dialogue* 42(2): 143–160.

Watts, M. 2015. "Adapting to the Anthropocene: Some reflections on development and climate in the West African Sahel." *Geographical Review* 53(3): 288–297.

7 Living with weeds

The dominant management approach to weeds[1] in Australia is seen at Bundanon, whose strategic plan puts this beautiful environment front and centre: 'The value of Bundanon's extraordinary, pristine, environment is self-evident to all who come here' (Bundanon Trust Strategic Plan 2011–2015).

Many visitors would agree with this statement, including artists, such as the Siteworks participants, who draw inspiration from Bundanon's landscapes. In contrast, others are horrified at the extent and intensity of weed infestation. Lantana (*Lantana camara*) lines the entrance road and blocks access to the river. Fireweed (*Senecio madagascariensis*) is scattered across the paddocks. So much so that the weed problem is considered by the strategic plan to be a significant threat to the Bundanon 'brand' (Bundanon Trust Strategic Plan 2011–2015, p. 5).

What are we to make of this contradiction between pristine nature and a 'brand-threatening' weed problem? Bundanon is anything but pristine, in the sense of being untouched by human hands and footprints. Instead we should understand it as being informed by the pristine ideal; an ideal of untouched nature. In this respect Bundanon is emblematic of a powerful but flawed current of Australian environmental thought; flawed in the sense that it is historically inaccurate (humans have been influencing the Australian landscape for tens of thousands of years), and of limited use in a future where human influences now dominate earth surface processes.

Bundanon and its companion property, Riversdale, are typically weedy places. Although we may want to focus on the pristine beauty of the sclerophyll forest, we can only do so by squinting. Fireweed and lantana are declared noxious weeds, and conventional management wisdom calls for their removal. It must be a headache for those whose core business is maintaining and fostering its artistic heritage. That heritage is partly one of colonialism, the sandstone farmhouse representing dispossession of Indigenous inhabitants as well as a lively cultural centre.

This chapter focuses on people's practical engagement with weeds, in both domestic and broader landscape contexts, in different parts of Australia. The process of living with – and killing – weeds can both rupture and reinforce people's understanding of the boundaries of nature. The Bundanon example shows the deep colonial heritage in the way we think about weeds – some have said this is a peculiarly Australian obsession. If that is the case, its intensity can throw light on the issues for other places. I suggest that we use weeds to help us understand the contradictions inherent in being human – in and of nature. Weeds, like humans, occupy ambiguous space in thinking about nature. We can use this ambiguity in our thinking. Different patterns of human ecological practice – growing crops, making roads, city blocks left derelict – encourage different combinations of weeds. Temporal boundaries based on past baselines interact with and intensify spatial bounding practices of belonging – they determine which spaces and species are marked as nature, native and considered to belong, and which are not. Weeds, gardens, roadsides and urban wastelands are becoming important parts of nature (Marris 2011).

The Bundanon Strategic Plan is not without challenge, such as that by Diego Bonetto, one of the contributing artists at Siteworks. His performance, 'Weeds R Us', saw the audience being served nettle soup and mallow tea. Bonetto expresses a more welcoming view of Australian nature than is provided for in most understandings of which plants belong here. His contribution to the Bundanon blog reflects his affectionate view of weeds:

> The strongholds of nature, the untameables, the unruly, the ones actually fighting back – blow after blow, seed after seed – our human prospective [sic] of what environment should look like . . . Nettle is everywhere, Urtica dioica, a blessed plant who fed human for millennia, and we will re appropriate this and more, mix it with native edibles and create what could be regarded as a cross cultural appreciation of the bounty nature has to offer . . .
>
> (Bonetto, 2010)

These are fighting words, confronting to most conventional understandings of Australian environmental management.

Human interactions with weeds vary hugely in space and time. Lantana is one of Australia's thirty-two declared Weeds of National Significance (WoNS). WoNS declarations provide a framework for managing (that is, killing) weeds across different jurisdictions and land tenures.[2] That peculiar Australian concept, Weeds of National Significance, causes great hilarity among my Swedish colleagues. Well, one explains, 'weeds are nature' in Sweden, and their national significance is understood differently. They are seen as plants growing where people do not want them, for example in a garden or in a crop, but they are understood to be part of nature. There is even a nature reserve specifically to preserve weeds. It is called Dalby Västermark, situated five kilometres east of Lund. Pesticides and artificial fertilisers are not used, making this a refuge of *en naturlig åkerogräsflora* (a natural field-weed flora),[3] in an area of Sweden's best arable land, so called 10+ quality. This is an example of *det biologiska kulturarvet* (biological cultural heritage), a concept promoted in recent years to include the biological remains of former land-use meadows, pollards and grazed forest. Hence weeds are part of the biological heritage associated with a long history of agricultural land use.

Despite warlike policy rhetoric to the contrary, it is impossible to get rid of all problematic weeds, and will be even more so under climate change scenarios. 'Living with' rather than 'completely removing' is only just emerging as a theme in the invasive species literature, where aspirations to the pristine have remained deeply embedded (Tassin & Kull 2015). The challenges of the Anthropocene call into question our governance framings of this and other environmental issues – a theme explored in detail in Chapter 8.

If the hope in this chapter were a colour, it would not be the deep green of environmental thought, but a greyer, more olive kind of green. We could even call it eucalyptus green, in deference to the tree whose evolutionary history has seen it dubbed the 'great Australian weed'. (And whose invasive status in many overseas countries is well established, however shocking to Australians.) Weeds help us understand the layers of human and natural history in a place and give us a sense of future processes. I argue that resilient, opportunistic, larrikin weeds may be more useful companions on the Anthropocene journey than we can yet imagine.

Before we get to the details of living with weeds, it is relevant to analyse how deeply implicated weeds are in the linear, managerialist narratives it will be necessary to disrupt. Their life or death is defined by temporal and spatial lines of belonging.

Belonging and nativeness

The critique of species nativeness in the social sciences sits within the wider discussion of the problematic boundaries around nature and culture addressed in earlier chapters. In this voluminous and complex literature, three particular trends are relevant to the themes of the book: 1) the place of humanness in relation to boundaries; 2) the influence of different temporalities; and 3) the implications for management of loosening or removing the conceptual boundaries.

Warren (2007) reviewed diverse ways in which nativeness and alienness have been used in relation to non-human species, emphasising the role of humanness in the definitions: 'In simple terms, native species are those which have autocolonised an area since a selected time in the past . . . and alien species are those which have been introduced by humans, intentionally or otherwise' (Warren 2007, p. 428, see also Preston 2009, p. 703). Warren also discussed the ways in which temporal and spatial contexts affected these definitions, for example the ways certain species might be considered native to Scotland. This question of definition illustrates an underlying ontological dilemma. As Lien and Davison argued, 'the biological classification of alien species . . . rests on an ontological distinction between human and non-human' (2010, p. 238).

The responses that Warren drew (Richardson et al. 2008, Preston 2009, Warren 2009) illustrate the ways in which such critique is contentious in the natural sciences. One example is the disagreement between Warren and Preston on the role of humanness in the definition. For Preston:

> The native/alien classification is one which distinguishes species on the basis of their dispersal to an area by human vectors; it does not make sense to apply

it to humans themselves . . . I find myself in agreement with Warren when he argues that the native/alien concept 'can only be applied if we exclude ourselves from it' but I do not regard this as something that 'destabilizes the alien/native framework'. It is merely a feature of this particular classification.

(2009, p. 708)

The two seem to agree on the value of this understanding for interpreting historical patterns of biogeographic change – the question of how things came to be the way they are, or description – rather than as a prescription for conservation management decisions (Warren 2009). Clark identified another very specific manifestation of separating out the human:

as globally oriented eco-activists, it is our task to exercise our own mobility and interactive capacities in order that we might find new ways to keep nature inactive and at home . . . we have not in the least ceased to be concerned with contamination, nor given up the patrolling of 'natural borders' or abandoned the rituals of purification.

(Clark 2002, p. 107)

Within biogeography and ecology, many discussions of nativeness have taken place in the context of invasion ecology, influenced by the broader context of the so-called new ecology, or non-equilibrium ecology (Wu & Loucks 1995, Stott 1998, Scoones 1999). Richardson (2011) provides a recent encapsulation of the field of invasion ecology, as its contributors include leading researchers and its chapters include a number of meta-analyses of recent literature.

In their chapter in the Richardson volume, Chew and Hamilton (2011) analyse historical conceptions of nativeness, developed by the British botanists John Henslow and H. C. Watson in the 1830s from common law concepts of native and alien, and by the Swiss phytogeographer Alphonse de Candolle in the 1850s. Chew and Hamilton show how these nineteenth-century conceptualisations provided the basis for contemporary understandings, with some modifications. Indeed they argue that 'it is remarkably easy to unravel the conception of biotic nativeness' (p. 44) from its foundation in pre-Darwinian ideas; that 'biotic nativeness is theoretically weak and internally inconsistent, allowing familiar human desires and expectations to be misconstrued as essential belonging relationships between biota, places and eras' (p. 36).

Dominant biogeographic schemes in use today 'share the tradition of distinguishing natives from non-natives by evidence of human intervention and a resulting range expansion' (Chew & Hamilton 2011, p. 40; see also Richardson et al. 2000, Pyšek et al. 2004). 'Human dispersal is said to render populations, and indeed any successor populations, non-native . . . Nativeness is therefore revocable, but non-nativeness is permanent. Being once human-dispersed accomplishes a mutagenic denaturing' (Chew & Hamilton 2011, p. 36). Spatial and temporal complexity in what 'human intervention' means results in many exceptions – the categorisation is suspended for livestock and crops, to whom are extended 'rights of occupancy'. Chew and Hamilton list many other exceptions:

> In summary, an olio of ideas from pre-Darwinian botany and pre-Victorian English common law still underpins even the most recent, expert conceptions of biotic nativeness. To the (wide) extent that biotic nativeness is considered actionable and presumed to rest on scientific findings, it is important for scientists, journalists, lawmakers, conservationists and other citizens to understand that those findings express some common beliefs about humans, but nothing about the essences of biota or of particular taxa.
> (2011, p. 40)

Concluding that the label 'native' is 'uninformative', 'deceptive', 'poorly founded' and 'hampering progress in ecological science', Chew and Hamilton have much in common with evolutionary ecologist Stephen Jay Gould's argument that:

> "native" plants cannot be deemed biologically best in any justifiable way . . . "Natives" are only the plants that happened to arrive first and be able to flourish . . . while their capacity for flourishing only indicates a status as "better than" others available, not as optimal or globally "best suited".
> (Gould 1997, p. 17; see also Sagoff 2003, Hattingh 2011)

Chew and Hamilton's argument throws down the gauntlet to the rest of the Richardson book, which seems to have accorded them a token presence. Some ecological writers consider that 'one cannot understand the trajectories of particular invasions by drawing a categorical distinction between introduced and native species. Rather, each species must be studied in its own right by examining how it interacts with other species during succession' (Simberloff 2011a, p. 20).

However, most other chapters proceed as if Chew and Hamilton's chapter had not been written. The book concludes with a glossary in which the unproblematised divide – on the basis of their human relations – is maintained. For example, *native species* are those 'that have evolved in a given area or that arrived there by natural means (through range expansion), without the intentional or accidental intervention of humans from an area where they are native' (Richardson et al. 2011, p. 416), and *alien species* are those 'whose presence in a region is attributable to human actions that enabled them to overcome fundamental biogeographical barriers (i.e. human-mediated extra-range dispersal)' (p. 410).

The issue continues to be controversial in the ecological sciences. Davis (2009, p. 166) has pointed out that there are a range of views within invasion biology about whether native species are inherently more valuable than others (see, for example, Brown & Sax 2004, 2005, cf. Cassey et al. 2005). When Davis et al. (2011) published an article titled 'Don't judge species on their origins', it was immediately controversial, generating heated discussion (e.g. Simberloff 2011b). The concept of novel ecosystems is similarly provocative (Murcia et al. 2014,

Hobbs et al. 2014). Invasive species are understood as part of the mosaic of anthromes. Ellis argues that anthromes are nevertheless not necessarily deleterious to natives, and that the 'majority of native species may be capable of maintaining viable populations in anthromes, at least over the near term' (2013, p. 370). (The inclusion of humans is what makes anthromes so radical against these other schema.) To reiterate the argument from Chapter 2, Hobbs attributes the intensity, indeed vitriol, of the debate to a grieving for the pristine.

A variety of empirical examples in the social sciences are now being assembled to further unsettle the native/non-native binary. Gibbs et al. (2015) show the contingent categorisation of camels in Australian history. In different contexts they have been both ferals that themselves need to be culled, and effective weed managers that selectively graze the invasive plant parkinsonia (*Parkinsonia aculeata*). Martin and Trigger (2015) examine the process of 'becoming indigenous' for three different trees in the Gulf of Carpentaria region of Australia's north. A coolabah (*Eucalyptus coolabah*), a date palm (*Phoenix dactylifera*) and a coconut tree (*Cocos nucifera*) each have diverse patterns of cultural associations and ways of belonging for human residents, different again from the ways in which ecologists would classify them.

If nativeness is not a robust concept in biological or ecological terms, its use as an axiom of management – supposedly founded on that science – is problematic. Nevertheless it continues to be a key category of management in Australian weed management. My colleagues and I have argued that the entanglement of 'nativeness' with other connected themes, for example the value of maintaining species diversity, or how to deal with problematic invasiveness, is hampering the management of those latter issues. There are many good reasons to continue to attempt to deal with invasive species, but the focus should be on the behaviours and effects of particular species and their interactions, rather than a pre-given status as native or non-native. This is increasingly recognised in the natural sciences literature (Pejchar & Mooney 2009, Schlaepfer et al. 2011, Davis et al. 2011). The effects of both natives and non-natives vary in both space and time, but it is not because of their nativeness or otherwise per se. In these 'novel ecosystems' (Hobbs et al. 2006), many organisms are changing their behaviour. Hobbs et al. have pointed out that a number of non-native species now constitute important components of ecosystems, 'for example, many butterfly species in California now depend on non-native plants for some or all of their food resources' (2009, p. 602). Non-native gorse in New Zealand provides important nursery plants and habitat for both plant and animal species (Barker 2008). In turn, native plants have changed their behaviour in these new associations (Ginn 2008). To the extent that humans can and should seek to intervene in managing ecosystems, we should focus on outcomes and processes rather than on the imposition of prescribed categories of being.

A further pragmatic concern is accelerating global environmental change and the impracticality – even if it were desirable – of restoration to a baseline state (Ginn 2008, Davis et al. 2011). Recognising that these decisions are inextricably entwined with cultural dispositions, Hobbs et al. (2010, p. 483) argue that the concept of 'naturalness' 'is no longer suitable as a management objective in park and wilderness areas' because it has so many different meanings, each of which can be contested. Decisions that distinguish between the positive and detrimental impacts of invasive species, 'will depend significantly on cultural values toward nativeness and exoticism and the ways in which such beliefs change in the coming decades' (Hobbs et al. 2009, p. 603), rather than being an outcome of science. Acknowledging the cultural bases of decisions will become even more important if, as argued by some, increasing levels of human intervention – in the form of translocation, for example – will be needed to save some species from climate change (Hoegh-Guldberg et al. 2008).

If the concept of biotic nativeness dissolves under empirical scrutiny in Australia as elsewhere, and is shown to be 'theoretically weak and internally inconsistent' (Chew & Hamilton 2011, p. 36), a number of aspects of our attempted management of Australian biota and landscapes need rethinking. In particular we need to think about how to create more open ecological futures (Staddon 2009) in a time of climate change. None of this discussion is to deny the significant impacts on abiotic and biotic life that came with European settlement – far from it. But living with the consequences of those changes requires us to be open to the contingencies of both the past and the future.

As discussed earlier, we can envisage a much wider suite of regenerative and co-productive human actions if they do not always have to be imagined as restoring past nativeness. Examples include bushcare activities, biodiversity conservation in agriculture and adaptive management more generally. In practice, vernacular acceptance of the new, recombinant Australian natures (Low 2002) is widespread. Garden research consistently shows that most urban Australians embrace an eclectic combination of species in their gardens, the most popular garden types including exotic plant species, either alone or in combination with natives (Zagorski et al. 2004; Trigger & Mulcock 2005, Head & Muir 2007). For some this is a spatialisation of belonging, with a tolerance extended in the city that would not be acceptable in the bush (Head & Muir 2006a, 2006b). So there is hope to be found if we look at what people are actually doing. It is to these practices that I now turn.

Practices of living with weeds

If Australia has experienced distinctive historical processes of entrenching the boundaries of nature and nativeness, it also has a distinctive heritage of destabilisation that is arguably starting to take vernacular expression. These more pragmatic approaches sit somewhat uncomfortably with the binary narratives dominant in environmental management discourses. The examples discussed here are extracted from studies of gardening practices in the temperate coastal cities of Sydney and Wollongong, and the desert town of Alice Springs, and professional weed management practices in the tropical rangelands of Australia's north.[4]

In the Australian context we know that many of our most problematic weeds have 'jumped the garden fence' (Groves et al. 2005). They were brought into the country for their beauty, and to remind migrants of home, but have behaved differently in Australian conditions, often becoming invasive. The garden focus is not because that will teach us how to deal with weeds in 'pure' nature somewhere else. Participants in the garden study were much more likely to talk about weeds in terms of their invasive qualities than their nativeness per se. Indeed for them a weed is in practice a plant that invades, or as one participant put it, weeds are plants that 'run over everything else'. This is a common sense understanding borne out of the labour of maintaining a garden, or frustration at not having the time or inclination to do so. In the garden project we traced the resilience of separationist approaches to the overlay of indigeneity/non-indigeneity atop other dualisms, and their rupture, to situations of close everyday engagement between people, plants, water and birds (Head & Muir 2006b).

Separationist approaches and the work of purification

The clear threshold of nativeness discussed above was shared by a group of gardeners identified in our backyard research as committed native gardeners or purists. Although they are a small minority, it is important to understand their practices and motivations, partly because they reflect the dominant ecological management discourse. (Albeit their practices are as hybrid as any other, since this view of nature requires exceptions to be made for dogs, cats and human selves.) Consistent with definitions of weeds discussed above, a clear divide between humans and nature was reinscribed in the way the human self was exempted from the category of invasive alien. A strong social separation is also seen when attempts at species purification intensify social boundaries with neighbours. Participants who were strongly committed to restoring native trees indigenous to their area were often in conflict with neighbours over the choice of plantings. Perhaps the most profound contradiction of the narratives of purity is that, although they are articulated in ways that exclude people, or in which people are invisible, any attempt to maintain or foster the dominance of locally indigenous species in a backyard requires an enormous amount of human effort, at least as much as maintaining a weed free and luxuriant lawn. It is not labour which can be invested just once, but must be ongoing if it is to be successful. This group tended to express disparaging attitudes towards 'exotic' or 'foreign' plants, and the neighbours who enjoyed them. For them, the importance of natives as 'belonging' was paramount. For some in this group the notion of gardening, with its connotations of planting and humanly assisted productivity, was inaccurate. Rather they saw themselves as restorers of native bushland, or eradicators of weeds that prevent native bushland restoring itself.

Claire in Alice Springs had been on a mission for many years to remove the introduced buffel grass (*Cenchrus ciliaris*) from her family's block:

> My object is to eradicate all the buffel and noxious grasses out and let the natural regrowth occur and it's just, it's my art; we have a sense that if I can achieve that to a high degree that that's my art within the community.

Soon after rain is an important time for the removal of buffel, when it rapidly responds by creating a green sward across the landscape. Claire described the practice of her art in very embodied terms. She told of regularly starting at 4.30 or 5.00 a.m. on summer mornings and removing buffel for three hours before the business of the day began, coming inside drenched in sweat and satisfied with her exertions. This physical, almost sensual, engagement and investment of labour was similar to that described by passionate gardeners who were planting things, and lawn mowers who were removing things.

Many gardeners, and not just purists, were active in weed removal beyond their fences, seeing different types of nature as belonging in different places. This is a further example of weeding as a *spatial* purification. Claire was not the only Alice Springs resident who removed buffel. Chris, who had an extremely manicured backyard including lawn and rose beds, was active in weed removal beyond her open mesh fence, seeing different types of nature as belonging in different places (cf. Gill et al. 2010). Gardeners who position most of the non-human world as belonging outside the city often had backyard zonations in which the proportions of 'native nature' increased with distance from the house. Dan expressed the ambivalence about their own belonging of many Euro-Australian gardeners:

> I guess the other angle is that you know ultimately we're probably not meant to be here either in terms of, you know, white Anglo-Saxon human beings. And then everything we eat, well 99 per cent of the things we eat aren't native to Australia either.
>
> (Dan, Port Kembla)

While he recognised this tension in the broader society, Dan and his partner Sue had a much more welcoming approach to exotic plants in their permaculture garden than did committed native gardeners. They described using known weeds to create animal habitat. They also had a declared noxious weed, water hyacinth (*Eichhornia crassipes*), in one of their ponds, trying to use it to remove heavy metals from the soil in their suburb close to heavy industry.

Although the native-exotic divide was expressed strongly by many workers in northern Australian pastoral and public land management, particularly those trained in the ecological sciences, purist approaches were barely seen in practice. Pragmatic approaches were no doubt due partly to the extensive areas involved and the enormity of the task. It is also the case in the northern Australian study that most people interviewed were working in a professional context rather than their own domestic environment. They may have felt less responsible for the overall outcome. There is a clear connection in both studies between the diverse everyday engagements in a more-than-human world (struggling with weeds, developing practical knowledge of how exotic and native species behave, enjoying birds) and the rupture of more separationist views of nature. It is important to emphasise that the garden and the savanna are not coincidental in these transformations. They should not be understood as separate field sites where we can observe the expression of pre-constituted attitudes and practices. Rather they are places of active making and remaking, of both humans and non-humans. This is not to say that learning to live with invasives is a comfortable process, nor one without hard decisions between different kinds of environmental damage, as the following sections illustrate.

Living with weeds often necessitates killing them

One circumstance we have not discussed in relation to grief is the lack thereof, specifically the lack of grief for the killing of plants. In the garden study the only deaths that were grieved for were trees. There seems to be something about the size, life form and longevity of trees that leads people to ascribe more rights to them than other forms of life. Barbara described the clearing they had to do when she and her husband came to their house as 'quite heart breaking, especially for Brian, cutting down trees [camphor laurels, *Cinnamomum camphora*] that were weeds'.

Our study on rubber vine (*Cryptostegia grandiflora*) in northern Australia provided a particularly adversarial example (Head et al. 2015). In that case people were fire-bombing rubber vine from a helicopter, at temperatures sufficient to ignite the latex in the plant – turning it against itself to kill the plant. In this example humans have altered their practices in order to become more effective killers, while trying not to kill too many other plants. Rubber vine is itself a killer of other plants and animals, strangling other plants and creating shelter for feral pigs and wallabies, which in turn displace smaller mammals such as bandicoots and pademelon.

The ethics of death and killing relate to the wider assemblages, including the humans involved in the process, and other plants, animals and insects that may be affected. This is particularly the case in the northern Australia context, where the sheer scale of weed management operations makes herbicide the main weapon of invasive plant management. Weed work is focused in the early wet season, as herbicide must be applied in the plants' main growing season to be effective. The heat and humidity of this season make such working conditions 'seriously oppressive' for the human participants (Atchison & Head 2013, p. 962). There is also a major question over the extent to which Indigenous rangers – for whom weed control is a major source of employment – are disproportionately at risk from chemical spraying (Head & Atchison 2015).

Plants as subjects, and the limits of human control

Invasives are thought of as exerting agency much more commonly than other plants; the agency of weeds is clear, but unwelcome. For example, in considering how backyard gardeners interact with weeds, an in-depth linguistic analysis of interview transcripts showed the conditions under which the agency of animals and plants was understood by our human participants (Thomson et al. 2006). Animals and plants were construed as Agents when perceived as the cause of an unwelcome process or else undergoing something 'natural'. In general in these interviews, humans are the agents of what goes on. Animals and plants undergo processes or else just 'are' or 'happen'. However, if plants, animals and the physical elements are involved in agency, then typically the impact of their agency is construed as negative. In this way they are held responsible for their action.

The fact that the inanimate world is construed as positive only when it 'behaves' itself is consistent with the frequent concern in the broader sample over matter (e.g. trees, weeds, mess) being 'out of place'. It is tied up with the gardeners' reasons or motivations behind their gardening practices, and their understanding of what belongs in different places. Similarly, invasives in northern Australia are read as a threat, as implicitly recognised through the need to legislate, both as part of a group (invasive plants) and as named individual species (WoNS). When acting as 'invasives', plants are understood to be not only mobile but aggressively so, marching across whole landscapes. As we see further in Chapter 8, this process is never completely within human control.

Planty bodies challenge the unit of engagement

Plants have different bodies, and this makes the unit of engagement an empirical question rather than a foregone conclusion. Garden trees, discussed above, provide one example. But trees start life as seeds, which are interacted with differently. It might be expected that vigilance towards seeds and seedlings is much more a feature of the domestic garden than the savanna lands of northern Australia, but weed management in the latter depends on surprisingly micro-scale observations. In the north of Western Australia, Jenny Atchison and I paid particular attention to the scale of the body – bodies of plants, animals and humans (Atchison & Head 2013). These are all monitored at strategic locations such as the border between Western Australia and the Northern Territory, where stock and tourist movements are both subject to surveillance. Seeds are stuck in the matted tails of horses, cattle and feral donkeys. They are picked up by socks, shoes, tents, caravans and human bodies. Stock inspectors described to us the way they rub horse and cattle tails between their fingers and hands, rolling them around to feel for seeds. They probe and monitor fresh dung. We argued that eradication – usually thought of as a separationist process, a process of pulling apart individual plants, and pulling apart plants and people – is rather a process of living, and dying, together. These bodies are not demarcated and separate, but are already and intimately in relation, albeit not always comfortably. Planty perspectives open up new ways of thinking about bodies and their boundaries. We are challenged to consider further the appropriate unit of ethical engagement – when and where should we worry about individuals, collectives, species? These governance challenges are discussed further in Chapter 8.

The labour of weeding – a job never finished

Weed management in gardens takes a lot of effort (whether pleasurable or not), invested over the long term. Initial investment and maintenance require different kinds of labour and vigilance. It is a job that will never be finished, but carefully targeted, long-term labour can make a difference. The most successful and contented garden weed managers are comfortable with the fact that plants have a life of their own and do their own thing. Like housework, there is always weeding to be done, as Quentin recounted:

> Mainly at the moment I would get out there and actually weed it and tidy it up again. It actually looked good at one time but I've just left it because I've been on shift work and just so busy lately that I haven't had time to actually do anything, so it's just gone on to how it has at the moment.
>
> (Quentin, Albion Park)

The labour of initial weed removal is different from maintenance. Various study participants described the huge task of removing lantana – including with goats and bulldozers – when they moved into their house and garden. The maintenance task is much more painstaking and requires ongoing, habitual commitment, as described by Jo the bush regenerator:

> Now where that fern is, all that area was all covered in madeira vine and lantana, and so I've just gradually got rid of it all and I have to keep vigilant. The madeira vine still comes back but we've just worked on, I just work at it, you know, it's painstaking work but I suppose I seem to have the right kind of mentality . . . when something gets away you just go back into one area and clean that completely . . .
>
> (Jo, northern Wollongong)

The issues that Jo was dealing with on an ongoing basis were very similar to those encountered by professional weed managers in more remote areas. For example, Dennis's experience, if somewhat more aggressive than Jo's, was very similar:

> I'm aware that I'm bashing my head against a brick wall. I killed millions yesterday. Millions more will grow in their place, but I can't wait to get back there and kill all them ones as well.
>
> Oh, I suppose you have to be philosophical about it and say . . . this job will go on forever whether it'll be me or someone else and in 200 years' time your descendant will be interviewing my descendant about weed control. They'll be different weeds but there will always be an issue with weeds I believe. You'll be managing them different but there will always be an issue with weeds.
>
> (Dennis, weed manager WA).

Dennis combined some aspects of native purism, expressed in his hatred for weeds, with both a pragmatic acknowledgement that he was never going to win, and a sheer stubbornness to keep going.

In the garden context, it was quite rare for people to say they enjoyed the work of weeding. Margaret of Campbelltown provided one example: 'I sometimes think weeding and tidying up is work. But once I get into it I really like it. I like puddling around in the dirt'. There was widespread recognition among garden study participants of the dynamism and constant change in the non-human world, including the view that nature has a life of its own beyond human control. While those who did not identify as gardeners – people who tended to be in pursuit of a 'low maintenance' garden – often aspired to a finished state, for gardeners, the notion that the garden could ever be 'finished' was incomprehensible.

Living with weeds influences and depends on inter-human conflict and cooperation

The conflict is not only the obvious one about adversarial relations with the plants, but also the kind of inter-human conflict that can be engendered. In the garden study, trees and lawn were subject to love-hate relations, and hence differing interpretations and experiences by their humans. Plants that transgressed the borders of private space could be considered invasive, particularly eucalypt trees that were frequently considered 'messy' because of the way they drop leaves and bark, and 'dangerous' because of the way they drop branches.

In other cases, weeds bring people together in less structured ways, often as neighbours. Where seeds are water distributed, strategic management practice is to work progressively downstream within a catchment so that downstream areas are not continually reinfected. This can be particularly difficult when weeds are the source of long-running conflicts between neighbours. In one case, a well-resourced (through grants) group at the downstream end of a catchment lamented that the work they were doing was almost useless because they were unable to work together with their upstream neighbours after other long-standing problems. In this case, the weeds were not the source of the problem, but they continued to add to the conflict with each seeding event. In contrast, coming together with neighbours can also be constructive. This is most likely to be the case if people have worked together previously or if they have achieved mutually beneficial results like winning land management grants. And as in the example Ben (weed manager, WA) describes below, these collaborations are easier when everyone is after the same outcome and well resourced:

> so you've really got five stakeholders and they've all got the same goal. L . . . Creek which runs down from the mine is just infested with bellyache bush and also a little bit of rubber vine there. It hasn't always been there. Traditionally they used to use that creek a lot for fishing. They can't now, the water quality's not that good and it's just full of weeds, you can't even get access to a lot of that creek system. So the environmental section of [large Mining Company], they have been providing funding for aerial burning of the weeds and the rangers work on that same section. They give us assistance there. Also the pastoralists, you've got all of them, even the traditional owners

> [Indigenous land owners], helping with the pre-season fire burning there as well. They've got a fairly good idea of what's the best time to burn and what does burn and doesn't burn. So that's been fairly good, that collaboration and all those stakeholders in that particular part.

Ben's quote illustrates the challenges of bringing together diverse stakeholders across a river catchment in which the distribution of weeds by water can negate preventative activities in other parts of the basin. This is just one example of how the governance of invasive plants is itself an extremely complex matter as lively plants pay no heed to spatial or jurisdictional boundaries.

Finally, I argue that we need a more open acknowledgement of the contradictions, edginess and difficult choices that attend contemporary Australia's engagements with nature. The times require us to go beyond the ideal of a pristine past and more honestly face a fraught, unpredictable and surprising future. In fact the wording has changed in Bundanon's more recent strategic plan (2015–2018). The equivalent section in the new plan reads:

> The value of Bundanon's spectacular landscape is self-evident to all who come here. The Trust will maintain its commitment to improving weed infested areas of the property, much of it now transformed through the Living Landscape partnership, and to opening up more areas for recreation and research.

This example illustrates one way in which framings can shift, as practitioners grapple with realities on the ground. I do not know whether the active critique and engagement fostered in Bundanon's arts/science partnerships also contributed, but it would be nice to think they did. Weeds may offer hope, as did the *Typha* in Chapter 6, in their capacity to be healers of broken landscapes (Crosby 2004, p. 170). Living with weeds does not necessarily mean giving up on, or 'lowering the bar' of good environmental practice (Murcia et al. 2014, p. 551). It is true, as Mather and Marshall (2011) noted in relation to animal biosecurity, that the notion of 'living with' can be vague and ambiguous; it is important to attend carefully to what it means. As we have seen here, gardeners and practitioners can be our instructors, as Chapter 8 considers in exploring the implications for governance.

Notes

1 The terms 'weed' and 'invasive plant' are used interchangeably in this chapter. Although I prefer 'invasive' as being more accurate to describe problematic plant behaviour, 'weed' is in common use in legislation, policy, institutional titles and the conversations of interview participants.
2 For further details see www.environment.gov.au/biodiversity/invasive/weeds/weeds/lists/wons.html. Last accessed 11 August 2015. The process of governing WoNS is discussed further in Chapter 8.
3 www.lansstyrelsen.se/skane/Sv/djur-och-natur/skyddad-natur/naturreservat/lund/dalby-Vastermark/Pages/_index.aspx. Last accessed June 2012. My thanks to Joachim Regnéll for translation.

4 A broader range of practices and challenges, drawing on additional studies, is discussed in Head et al. (in press). Here I concentrate on comparative insights from two different research contexts in which I have worked. For details of the garden research, including methods, see Head and Muir (2006b, 2007) and Head (2014). For the tropical rangelands research see Atchison and Head (2013), Head et al. (2015) and Head and Atchison (2015). The backyard project was a large study of 265 backyard gardens and their 330 owners (some couples and family groups participated). Most of the fieldwork was carried out between 2002 and 2005, a period of intensifying drought in southeastern Australia. The sample population comprised 122 backyards in Sydney, 119 in Wollongong and 24 in Alice Springs, in areas that spanned the socioeconomic, demographic and ecological variability of each city area. The rangelands research encompassed documentary and policy analysis, semi-structured interviews with fifty-one participants in different contexts (policy-makers, scientists, landholders, ranger groups, landcare volunteers), and participant observation of field activities and stakeholder workshops, across Queensland, the Northern Territory and Western Australia, between 2010 and 2013.

References

Atchison, J. and L. Head. 2013. "Eradicating bodies in invasive plant management." *Environment and Planning D: Society and Space* 31: 951–968.

Barker, K. 2008. "Flexible boundaries in biosecurity: Accommodating gorse in *Aotearoa* New Zealand." *Environment and Planning A* 40(7): 1598–1614.

Bonetto, D. 2010. "How to cook nettles." *Bundanon Siteworks – The Blog*. Accessed 11 August 2015. Available at https://siteworksblog.wordpress.com/2010/09/14/on-how-to-cook-nettles/.

Brown, J. H. and D. F. Sax. 2004. "An essay on some topics concerning invasive species." *Austral Ecology* 29(5): 530–536.

Brown, J. H. and D. F. Sax. 2005. "Biological invasion and scientific objectivity: Reply to Cassey et al. (2005)." *Austral Ecology* 30(4): 481–483.

Bundanon Trust Strategic Plan 2011–2015. Available at www.bundanon.com.au.

Bundanon Trust Strategic Plan 2015–2018. Available at www.bundanon.com.au.

Cassey P., T. M. Blackburn, R. P. Duncan and S. L. Chown. 2005. "Concerning invasive species: Reply to Brown and Sax." *Austral Ecology* 30(4): 475–480.

Chew, M. K. and A. L. Hamilton. 2011. "The rise and fall of biotic nativeness: A historical perspective." In *Fifty Years of Invasion Ecology: The Legacy of Charles Elton*, edited by D. M. Richardson, 35–48. Oxford: Blackwell.

Clark, N. 2002. "The demon-seed: Bioinvasion as the unsettling of environmental cosmopolitanism." *Theory, Culture and Society* 19: 101–125.

Crosby, A. W. 2004. *Ecological Imperialism: The Biological Expansion of Europe, 900–1900*. 2nd ed. Cambridge: Cambridge University Press.

Davis, M. A. 2009. *Invasion Biology*. Oxford: Oxford University Press.

Davis, M., M. Chew, R. Hobbs, A. Lugo, J. Ewel, G. Vermeij, J. Brown, M. Rosenzweig, M. Gardener, S. Carroll, K. Thompson, S. Pickett, J. Stromberg, P. Del Tredici, K. Suding, J. Ehrenfeld, J. Grime, J. Mascaro and J. Briggs. 2011. "Don't judge species on their origins." *Nature* 474: 153–154.

Ellis, E. C. 2013. "Sustaining biodiversity and people in the world's anthropogenic biomes." *Current Opinion in Environmental Sustainability* 5: 368–372.

Gibbs, L., J. Atchison and I. Macfarlane. 2015. "Camel country: Assemblage, belonging and scale in invasive species geographies." *Geoforum* 58, 56–67.

Gill, N., P. Klepeis and L. Chisholm. 2010. "Stewardship among lifestyle oriented rural landowners." *Journal of Environmental Planning and Management* 53(3): 317–334.

Ginn, F. 2008. "Extension, subversion, containment: Eco-nationalism and (post)colonial nature in Aotearoa New Zealand." *Transactions of the Institute of British Geographers* 33(3): 335–353.

Gould, S. J. 1997. "An evolutionary perspective on strengths, fallacies, and confusions in the concept of native plants." In *Nature and Ideology: Natural Garden Design in the Twentieth Century*, edited by J. Wolschke-Bulmahn, 11–19. Washington: Dumbarton Oaks.

Groves, R. H., R. Boden and W. M. Lonsdale. 2005. *Jumping the Garden Fence: Invasive Garden Plants in Australia and Their Environmental and Agricultural Impacts.* CSIRO report prepared for WWF-Australia, Sydney.

Hattingh, J. 2011. "Conceptual clarity, scientific rigour and 'The Stories We Are': Engaging with two challenges to the objectivity of invasion biology." In *Fifty Years of Invasion Ecology: The Legacy of Charles Elton*, edited by D. M. Richardson, 359–376. Oxford: Blackwell.

Head, L. 2014 "Living in a weedy future: Insights from the garden." In *Rethinking Invasion Ecologies from the Environmental Humanities,* edited by J. Frawley and I. McCalman, 87–99. London: Earthscan for Routledge.

Head, L. and J. Atchison. 2015. "Entangled invasive lives: Indigenous invasive plant management in northern Australia." *Geografiska Annaler: Series B, Human Geography* 97(2): 169–182.

Head, L., J. Atchison and C. Phillips. 2015. "The distinctive capacities of plants: Rethinking difference via invasive species." *Transactions Institute of British Geographers* 40(3): 399–413.

Head, L., B. M. H. Larson, R. Hobbs, J. Atchison, N. Gill, C. Kull and H. Rangan. In press. "Living with invasive plants in the Anthropocene: The importance of understanding practice and experience." *Conservation and Society.*

Head, L. and P. Muir. 2006a. "Edges of connection: Reconceptualising the human role in urban biogeography." *Australian Geographer* 37(1): 87–101.

Head, L. and P. Muir. 2006b. "Suburban life and the boundaries of nature: Resilience and rupture in Australian backyard gardens." *Transactions of the Institute of British Geographers* 31(4): 505–524.

Head, L. and P. Muir. 2007. *Backyard: Nature and Culture in Suburban Australia.* Wollongong, NSW: University of Wollongong Press.

Hobbs, R. J., S. Arico, J. Aronson, J. S. Baron, P. Bridgewater, V. A. Cramer, P. R. Epstein, J. J. Ewel, C. A. Klink, A. E. Lugo, D. Norton, D. Ojima, D. M. Richardson, E. W. Sanderson, F. Valladares, M. Vilà, R. Zamora and M. Zobel. 2006. "Novel ecosystems: Theoretical and management aspects of the new ecological world order." *Global Ecology and Biogeography* 15: 1–7.

Hobbs, R. J., D. N. Cole, L. Yung, E. S. Zavaleta, G. H. Aplet, F. S. Chapin III, P. B. Landres, D. J. Parsons, N. L. Stephenson, P. S. White, D. M. Graber, E. S. Higgs, C. I. Millar, J. M. Randall, K. A. Tonnessen and S. Woodley. 2010. "Guiding concepts for park and wilderness stewardship in an era of global environmental change." *Frontiers in Ecology and the Environment* 8(9): 483–490.

Hobbs, R. J., E. Higgs and J. A. Harris. 2009. "Novel ecosystems: Implications for conservation and restoration." *Trends in Ecology and Evolution* 24(11): 599–605.

Hobbs, R. J., E. S. Higgs and J. A. Harris. 2014. "Novel ecosystems: Concept or inconvenient reality? A response to Murcia et al." *Trends in Ecology and Evolution* 29(12): 645–646.

Hoegh-Guldberg, O., L. Hughes, S. McIntyre, D. B. Lindenmayer, C. Parmesan, H. P. Possingham and C. D. Thomas. 2008. "Assisted colonization and rapid climate change." *Science* 321: 345–346.

Lien, M. E. and A. Davison. 2010. "Roots, rupture and remembrance: The Tasmanian lives of the Monterey Pine." *Journal of Material Culture* 15(2): 233–253.

Low, T. 2002. *The New Nature: Winners and Losers in Wild Australia.* Camberwell, Australia: Viking.

Marris, E. 2011. *Rambunctious Garden: Saving Nature in a Post-Wild World.* New York: Bloomsbury.

Martin, R. J. and D. Trigger. 2015. "Negotiating belonging: Plants, people, and indigeneity in northern Australia." *Journal of the Royal Anthropological Institute* 21(2): 276–295.

Mather, C. and A. Marshall. 2011. "Living with disease? Biosecurity and avian influenza in ostriches." *Agriculture and Human Values* 28(2): 153–165.

Murcia, C., J. Aronson, G. H. Kattan, D. Moreno-Mateos, K. Dixon and D. Simberloff. 2014. "A critique of the 'novel ecosystem' concept." *Trends in Ecology & Evolution* 29(10): 548–553.

Pejchar, L. and H. A. Mooney. 2009. "Invasive species, ecosystem services and human well-being." *Trends in Ecology and Evolution* 24(9): 497–504.

Preston, C. D. 2009. "The terms 'native' and 'alien' — a biogeographical perspective." *Progress in Human Geography* 33(5): 702–711.

Pyšek, P., D. M. Richardson, M. Rejmánek, G. L. Webster, M. Williamson and J. Kirschner. 2004. "Alien plants in checklists and floras: Towards better communication between taxonomists and ecologists." *Taxon* 53(1): 131–143.

Richardson, D. M. 2011. *Fifty Years of Invasion Ecology: The Legacy of Charles Elton.* Oxford: Blackwell.

Richardson, D. M., P. Pyšek and J. T. Carlton. 2011. "A compendium of essential concepts and terminology in invasion ecology." In *Fifty Years of Invasion Ecology. The Legacy of Charles Elton*, edited by D. M. Richardson, 409–420. Oxford: Blackwell.

Richardson, D. M., P. Pyšek, M. Rejmánek, M. G. Barbour, F. D. Panetta and C. J. West. 2000. "Naturalization and invasion of alien plants: Concepts and definitions." *Diversity and Distributions* 6(2): 93–107.

Richardson, D. M., P. Pyšek, D. Simberloff, M. Rejmánek and A. D. Mader. 2008. "Biological invasions – the widening debate: A response to Charles Warren." *Progress in Human Geography* 32(2): 295–298.

Sagoff, M. 2003. "The plaza and the pendulum: Two concepts of ecological science." *Biology and Philosophy* 18(4): 529–552.

Schlaepfer, M. A., D. F. Sax and J. D. Olden. 2011. "The potential conservation value of non-native species." *Conservation Biology* 25(3): 428–437.

Scoones, I. 1999. "New ecology and the social sciences: What prospects for a fruitful engagement?" *Annual Review of Anthropology* 28: 479–507.

Simberloff, D. 2011a. "Charles Elton: Neither founder nor siren, but prophet." In *Fifty Years of Invasion Ecology: The Legacy of Charles Elton*, edited by D. M. Richardson, 11–24. Oxford: Blackwell.

Simberloff, D. 2011b. "Non-natives: 141 scientists object." *Nature* 475: 36.

Staddon, C. 2009. "The complicity of trees: The socionatural field of/for tree theft in Bulgaria." *Slavic Review* 68(1): 70–94.

Stott, P. 1998. "Biogeography and ecology in crisis: The urgent need for a new metalanguage." *Journal of Biogeography* 25: 1–2.

Tassin, J. and C. A. Kull. 2015. "Facing the broader dimensions of biological invasions." *Land Use Policy* 42: 165–169.

Thomson, E. A., C. Cleirigh, L. Head and P. Muir. 2006. "Gardener's talk: A linguistic study of relationships between environmental attitudes, beliefs and practices." *Linguistics and the Human Sciences* 2(3): 425–460.

Trigger, D. and J. Mulcock. 2005. "Native vs exotic: Cultural discourses about flora, fauna and belonging in Australia." In *Sustainable Planning and Development: The Sustainable World*, edited by A. Kungolos, C. Brebbia and E. Beriatos, 1301–1310. Southampton: Wessex Institute of Technology Press.

Warren, C. R. 2007. "Perspectives on the 'alien' versus 'native' species debate: A critique of concepts, language and practice." *Progress in Human Geography* 31(4): 427–446.

Warren, C. R. 2009. "Using the native/alien classification for description not prescription: A response to Christopher Preston." *Progress in Human Geography* 33(5): 711–713.

Wu, J. and O. L. Loucks. 1995. "From balance of nature to hierarchical patch dynamics: A paradigm shift in ecology." *Quarterly Review of Biology* 70(4): 439–466.

Zagorski, T., J. B. Kirkpatrick and E. Stratford. 2004. "Gardens and the bush: Gardeners' attitudes, garden types and invasives." *Australian Geographical Studies* 42(2): 207–220.

8 Governing the ungovernable?[1]

Smoke is a common sight when you fly into Darwin in the dry season. Pastoralists, Aboriginal communities and government fire managers use landscape burning for a variety of preventive and proactive reasons. They aim to reduce fuel loads to reduce the intensity of late-dry-season fires, and stimulate new growth of favoured species. Particularly dry and windy conditions in August 2011 led to the Territory's first catastrophic fire-danger warning (Brennan 2011).

A key contributor to those increased fuel loads was the invasive gamba grass (*Andropogon gayanus*), which grows to four metres tall, a common sight in the peri-urban bushland fringes around Darwin and Palmerston. Setterfield et al. (2013) calculated that fire management costs had multiplied about nine times over the previous decade, primarily due to gamba invasion. Gamba's main threat to biodiversity is its potential to radically alter vegetation structure in the ecosystems it invades, by promoting hotter more intense fires which reach up into the tree canopy and often result in tree death. Across northern Australia, the feared outcome is the transformation of the biodiversity-rich savanna into grassland (Rossiter et al. 2003).

In this chapter I use the example of gamba grass to discuss the issues of governance and management in a context of uncertainty and ongoing change. Invasive species are considered to be one of the most significant environmental problems of the twenty-first century, becoming more severe under climate change (McGeoch et al. 2010; Driscoll et al. 2012; Tassin & Kull 2015). In the gamba example the challenges of invasive plants, fire and climate change come together. Climate change and invasive species are now interacting to 'generally increase the risk and intensity of fire' (Settele et al. 2014, p. 289). Climate change may also decouple ecological relationships on which invasive species management through biocontrol currently depends (Settele et al. 2014).

As we discussed in Chapter 3, the uncertainty and non-linear changes of the Anthropocene confront land use policy developed for conditions of predictability and fixity. Surprise shifts and potential catastrophe 'challenge the steering capacity of governance at all political levels' (Duit & Galaz 2008, p. 311). The question of what it means to 'manage' or 'govern' an environmental or land use issue must be rethought. How are we to act if the total context may be essentially ungovernable? The chapter addresses this question and two others: 1) how well suited is a linear and instrumentalist policy process to more mobile and distributed agencies, causations and changes – even when well coordinated across spatial scales? and 2) specifically, what does it mean to live with invasive plants in ways that do not abandon all human responsibility? This research has been collaborative with Jenny Atchison (Head & Atchison 2015).

Scientific assessments of the reasons for failure to control invasive species draw attention to issues such as:

- inadequate policy,
- lack of funding,
- gaps in scientific knowledge,
- the complexity of managing across complex and increasingly fragmented land tenure landscapes (Simberloff et al. 2005; Epanchin-Niell et al. 2010; Driscoll et al. 2012).

The overall tenor of most invasive species management literature is that the problem can be fixed if the right combination of science, policy and funding is directed towards the problematic species. We contest that view, arguing that a more risky, unstable, experimental process is always in play. Atchison (2015) shows how the science of biocontrol is itself an experimental practice, and a correlation between science and control cannot be assumed.

A pasture grass introduced through government policies to assist the pastoral industry, gamba grass is (in principle) one of the most 'governed' species in Australia, if by governing we refer to the legislation and policy that surrounds it. By attending to the practices of governance as well as to the plant itself as a key part of the governance assemblage, we argue for a broader understanding of what contributes to, and constitutes, governance. By illustrating the practices of living with gamba, extending ideas discussed in Chapter 7, we suggest how policy and governance might be improved and reframed for a more enduring relationship.

Environmental governance and the challenges of invasive plants

A number of processes interact to create a challenging management and policy environment for invasive species, particularly plants. First, there is widespread recognition that invasive species interact with a diversity of other things in the landscape, including public perceptions and preferences, land use patterns and changes, fire and climate change. Second, governance – which takes place formally at national, state/territory and local levels in Australia – draws attention to the relationship between different scales, and the challenges of governing across scales from the international to the individual landholder.

Effectively responding to these challenges means acknowledging complexity. The environmental governance literature offers useful insights here in the attention it gives to the assemblages being governed. Rather than a linear relationship from science to policy to the problematic species, there is recognition of a less linear, complex network of actors and processes (Ogden et al. 2013). Indeed, researchers draw attention to failure, as an inherent and fundamental aspect of governing, and a constitutive part of regulation itself (Higgins 2004, Law 2006, Enticott 2014). They also draw explicit attention to the influence of non-human actants, for example, Lockie's (2004) analysis of Landcare organisations in a field characterised by distributed rather than unidirectional agency, and 'assemblages of governing' (Lockie & Higgins 2007). Connecting to our discussion in Chapter 7, we need to consider what we are governing – species, collectives, seeds? As Barker argues:

> For invasive plants and plant diseases, the inherent unpredictability of relational life leaves national and international governing bodies and international trade agreements scrambling to keep up, and poses problems for risk assessments based on current observations – as good behavior in one environment fails to guarantee docility in another – particularly in the context of climatic shifts.
>
> (2014, pp. 3–4)

As we saw in Chapter 7, attending to the *practices* of governing provides illumination. Everyday, fine-grained activities make up the broader assemblage that comes to be called governance. Many kinds of labour are involved in governing invasive plants even before it comes to the question of whether to kill them with chemicals or by hand; writing funding applications, reporting on funded grants, establishing and running community education programs, negotiating among adjacent landholders and undertaking occupational health and safety training. The labour of science involves 'unexpectedly intimate' practices between human bodies and plants (Atchison 2015).

In focusing here on the practices of invasive plant management and governance, it is not our intention to advocate 'anything goes', nor to give up on the very severe socio-ecological problems posed by invasives. Rather, it is to take issue with the assumption that all problems are there to be fixed and are fixable through 'good' policy and governance. The challenges of rapid and unpredictable change in the Anthropocene are overlaid on an existing situation in which contest, chaos and failure are a normal part of the policy and governance process (Higgins 2004, Mann & Absher 2014).

Invasive plant governance in Australia and the case of gamba grass

Gamba grass was introduced from Africa to Australia as part of a Commonwealth Government pasture improvement programme in the 1930s (Cook & Dias 2006). In its present genetic expression gamba can be understood as an Australian cultivar, a result of the Commonwealth Scientific and Industrial Research Organisation (CSIRO) experimentation and pasture release pathway (Anon. 1987). A perennial tussock growing to four metres, gamba grass produces large amounts of seed (70,000 per square metre) and grows in a wide range of environments (Flores et al. 2005). It out-competes native grass species and thus alters the composition, hydrological function and nutrient composition of grassland species (Rossiter-Rachor et al. 2009). Gamba grass now covers extensive areas in the Northern Territory and Queensland, and a number of incursions have been recently recorded in the north of Western Australia.

Because the three levels of government in Australia have different environmental responsibilities, invasive plant management is necessarily a cross-scalar process. The federal government has legislative responsibility for quarantine and prevention processes that attempt to prevent non-native species coming into the country, and also for biodiversity conservation through the Environmental Protection and Biodiversity Conservation Act 1999 (Cth) (EPBC Act). Land management generally is a state or territory responsibility, and most planning jurisdiction resides at the state and/or local government level.

In the view of Bulkeley et al. (2007), the practices of governance in Australia coalesce around *species* as the primary entity to be governed. A spatial prioritisation process operates underneath that, for each species. Until recently, gamba grass has been classified differently in state and territory jurisdictions, the primary focus being its declaration, and accompanying statutory management plan, in the Northern Territory in 2010. More recently, invasive plants have been examined by the Commonwealth under the 'key threatening process' provision of the EPBC Act. The first plant to be considered, gamba was listed as a threatening process in its own right under this mechanism (effective 2009), although a subsequent review now groups gamba together with five other introduced grass species (Beeton n.d.). Listing as a key threatening process triggers assessment for a national Threat Abatement Plan, which aims to provide the framework for 'consistency of management across all jurisdictions' (Beeton n.d., p. 14).

In a parallel process, a concerted effort has been made by the Commonwealth to provide more effective coordination and collective effort across jurisdictions. A crucial feature has been the development of the list of Weeds of National Significance (WoNS), which arose out of the 'National Weeds Strategy' (1997), re-released in 2007 as the 'Australian Weeds Strategy' (NRMMC 2007). Criteria against which WoNS species are assessed include factors such as invasiveness, impacts and the potential for spread. One of the main practical criteria is their governability; 'consideration was also given to their ability to be successfully managed'.[2] Although the federal government does not have legislative jurisdiction here, through this mechanism it can effectively direct weed management by channelling funding towards action on national priority WoNS species.

Although the Threat Abatement Plan is yet to be finalised, its combination with the WoNS strategy and the existing state and territory legislation arguably constitutes the most complex and rigorous governance framework for any plant in the country. Yet even in this example, a level of acknowledged ungovernability operates in the spatial prioritisation process beneath the species level plans. The

Northern Territory gamba grass management plan permits different kinds of practices in different management zones, based on where regulators consider certain weed outcomes are feasible or not. Classes are designated A (to be eradicated), B (growth and spread to be controlled) and C (not to be introduced in the Northern Territory), with mixed classes reflecting varying feasibility of control. In the AC zone, for example, eradication is considered feasible and cost effective. The overarching rationale in the BC zone is that the weed is only controllable, not eradicable. Such contingencies change over time; in the most recent review (DLRM 2014), eradication zones have been expanded. Different land tenures add further complexity. Some Indigenous lands, for example, rank amongst the least weedy (Franklin et al. 2008, Preece et al. 2010), but communities have highly variable socioeconomic capacities to manage weeds on their land. Properties returned from the Crown to native title holders are often in very poor condition, and there is a lack of clarity around who is obliged to manage weeds on native title lands (Duff & Weir 2013).

Weed policy is often framed as something that can be 'made clear' or fixed when good science, coordination and capacity building come together in the right way (AWC 2013). The problems of bringing such governance into being are recognised to a limited extent within policy. For example, the gamba grass strategic plan (AWC 2013, p. 3) views management as a 'shared responsibility' requiring 'coordination' and 'capacity'. That is, policy recognises that a particular kind of distributed human agency is necessary, in that many different kinds of labour and human effort are required even though legal responsibility lies with landholders. It must also respond to a clearly distributed and mobile plant agency, something commonly responded to through calls for better mapping and/or monitoring processes. Our contention is that a more realistic framing of weed governance, away from control, would present it as realised with always imperfect knowledge, across complex social and ecological landscapes, in a context where both humans and non-humans are active agents.

The practice of governing and managing invasive plants

Atchison and I conducted over fifty semi-structured interviews across northern Australia, between 2010 and 2013. The time period of fieldwork coincided with the declaration of gamba grass under the Weed Management Act NT in 2010, the establishment of formal governance processes and the raised profile of gamba due to its fire threat across the Northern Territory community. This section is structured around three principal aspects in which the practices of bringing governance into being are expressed most clearly:

- priority setting and species declaration,
- managing interacting risks – weeds and fire,
- responsibility.

Interview quotes are used in these sections as exemplars of wider trends.

Practices of priority setting and species declaration

Most managers agreed there are simply too many weeds to do something about all of them. In the Northern Territory, for example, managers assessed over ninety-seven species as part of their assessment. Increasingly, priority setting takes place via a risk assessment process which considers the risks to industry, the environment and human health, and the feasibility of control. According to government manager participants, priority setting must be 'formalised, transparent, evidence-based, and defensible', meaning that choices have to be made about how limited weed budgets will be allocated, and whether the legal ramifications of those choices will stand scrutiny in a court of law. But there is also a more complex politics of priority setting illustrated by the following interview extract:

> we won't be recommending any declarations because [some species are] too valuable and well established, and part of the risk assessment process is feasibility of control . . . Many of these things, the feasibility of control is so low now that if you plant it, it will spread so far – like buffel grass in the Centre [Central Australia]. Regularly there are calls for a declaration or discussion of making a weed of national significance, but it's not done because the industry resistance would be so high we probably couldn't get it done. And even if we could it would damage the credibility of the weed management agencies and make our jobs more difficult, make it more difficult to declare things in the future, and besides, what would declaration achieve? The feasibility of control is very low indeed.
>
> (Government Senior Manager A)

In this quote, a senior bureaucrat identifies two influential factors that could stop a damaging weed being declared; the ecological unfeasibility of stopping the spread of weeds already considered out of control, and the political opposition by groups such as pastoralists who value some weeds for economic purposes. Gamba grass is one example of the latter; many pastoralists have opposed its declaration as a weed because they still use it as a resource. To quote one Landcare coordinator describing the perspective of a pastoralist he knows: 'I'm actually shooting myself in the foot by spending time and effort spraying something that the cattle will eat to be replaced by something that they don't eat'.

This land manager and others describe without irony a risk assessment process in which 'feasibility' is a key variable. They refer to documents that outline the scientific risks, but the political risks and feasibilities they are explicitly juggling are just as important in the process. Particular industry groups are known to be very vocal and powerful lobbyists within consultative forums. By its very nature the process of species declaration acknowledges that some weeds cannot be managed, either being too widespread and costly to control, or too valuable to industry.

So what tipped the balance to getting gamba grass declared? The threats that it poses to biodiversity are relatively long term, and not particularly visible to the non-expert land manager. Its fire risk is more visible, immediate and a threat to human settlements and safety. The plant itself is imposing, clearly carrying a significant fuel load of dry matter in linear stands along fences and roadsides. Property and infrastructure loss in the Darwin area as a result of more intense gamba fires in recent years (Setterfield et al. 2013) has served to highlight the consequences of inaction. Fire risk has been crucial in mobilising political action, as explained by Government Senior Manager A:

> The reason we got gamba grass declared was because of its fire risk ultimately, because you know the Darwin rural area's got thousands of 10 hectare blocks so there's a lot of people out there and apart from the fact that it [gamba] likes bush blocks, it's ruining them, it's very much a fire risk and so it was politically possible to declare it . . . It is the worst of the terrestrial environmental weeds but there's plenty of others.

For those species that do end up at the top of the priority list, a key policy aim is to shift practice. Although there are in fact 'quite draconian powers' provided for in the Act (e.g. the right for weed officers to enter and search property), our government participants emphasised that enforcement was not something they were pursuing, and that prosecutions would need to be limited and strategic. In fact, there is probably limited social license to enforce the Act in the wider political context of the Northern Territory, where individual freedom and frontier independence are strong cultural values.

In this context, managers understand legislation and regulation as working to provide the incentives and lubricant for the flows of money and people which are required to make changing practices possible, as Government Officer A explains:

> It was really good to try and come from a position where, only two years before, the plant hadn't even been declared, to a complete change in legislative status and immediate need to involve everybody. So all major stakeholders, Power and Water, you know, the big players like Northern Territory Parks and right down to the 5 acre blockies. So we actually developed an assistance program [in] which we went out with spray equipment. We provided herbicide, we provided identification advice, we helped with property management planning, we developed a whole range of extension material . . . We did everything that we could think of to try and raise the profile of gamba and the damage that it was causing, and it actually had quite good results. It was nice to see.

The provision of herbicide that this officer notes raises the important issue that much invasive plant management across the extensive areas involved in northern Australia is dependent on the application of chemical herbicides. Funding is crucial because of the labour and equipment involved in their application.

In summary, examination of the practices of weed prioritisation and declaration brings to light a cultural politics of which the legislative and policy aspects are but one dimension. The process of weed governance itself acknowledges that the majority of weeds cannot be governed. Weeds exceed and burst out of governance in a variety of ways: when 'the horse has bolted' and control is practically not feasible, when they are so valuable to some human groups that political control is not possible, in spaces where containment rather than eradication is the goal and in political contexts when strong legislative powers exist but lack the social license to operate. On the other hand, the more visible risk of fire can mobilise necessary political action; the interaction of policies relating to different aspects of land use produces results with implications for weeds.

Practices of managing interacting risks – weeds and fire

The threat of intense fire as a consequence of gamba fuel loads now means that fire, rather than other ecological issues, sets the ongoing management priorities for some of the land managers we spoke to. The rapidly changing fire environment has consequences for slower-moving weed policy. One main principle of weed policy is to treat outlying infestations and work back towards the centre (March, 2011). Fire policy tends to the opposite spatial prioritisation; it necessarily concentrates on large central infestations where fuel loads are highest, those around human settlements and assets that need protection.

One government officer we spoke to is responsible for weed management on extensive tenure across the Northern Territory. When asked how she prioritised work on such a scale, this manager recognised the need to manage outlying patches or infestations as 'good' weed management. Yet in practice the threat of fire in the urban and peri-urban areas is a more immediate concern and takes up much of her time, as she explains:

> Gamba's our huge issue because of the fire risk. So yeah, we just have to prioritise where we're going to be working and spending the money . . . more in residential areas as well, if there's going to be a fire risk. Yes, I was actually checking a fire break, because we've got Territory Day on Monday, where people can let off fireworks, so fire's a huge issue . . . We've got a good eight metre break.
>
> (Government Officer B)

Stretched and competing resources are a common problem for government land managers. The workload of managing the patchwork fabric of this peri-urban area and its abundant edge effects means that some managers, like the one above, must more often prioritise the immediate risks of fire. Other land managers are responding to the fire threat posed by gamba grass by incorporating it into routine management activities. In this respect they are responding to both fire and weed regulations, albeit in different ways. Routine weed and fire management is more commonly encountered where weeds are a minor problem or where there is a legal obligation for weed removal. For example, one extractive industries member sprays and burns his mine site a couple of times a year.

Interviewee: The weeds don't affect us. It's just, you know, it's the right thing to do, I suppose, but it doesn't affect the operation. It doesn't really affect the operation at all. Yeah. It's more just an obligation, I suppose, under the [weed] regulations.

Interviewer: And do you have to do fire control work around your site as well?

Interviewee: We get the local volunteer bushfire brigade to do that, because we're reasonably close to a residential area. They like to ensure that all that's under control, and fire breaks are in, and they've done back burns before.

(Extractive Industry Member)

In this case managing gamba grass is now routine because managing weeds is outsourced to a contractor, and a minor imposition on the business. Managing fuel loads and fire threat is also a legal obligation and is in practice made routine through the 'free' provision of labour from the local fire brigade, as well as through their expertise in managing the risks of lighting fire breaks on particular days. Regular spraying of weeds works in combination with fuel reduction, reinforcing the routine. Managing weed risk and fire risk together is possible because both are outsourced and predictable; actions can be planned and scheduled. Hence the risks of spraying and of setting the fire breaks become someone else's expertise and their 'responsibility'.

In other situations managers have incorporated weed management work into routines where there is a clear business case, as well as legal obligations, to do so. An environment manager with a large air transport industry described their operations across the Northern Territory, recounting the risk assessment process in some detail. In highly regulated spaces there are multiple regulations to consider, and multiple 'keepers' to 'keep me on my toes'. Any fire is a significant attractant to black kites (*Milvus migrans*), which pose a serious bird-strike risk to aircraft. This manager spoke to us of 'heart palpitations' every time she sees a fire burning in and around the airport. Not only is there a bird-strike risk, but the ash fallout also poses a fire risk to underground fuel tanks and to refuelling activities. In this situation, she manages fuel loads by slashing and spraying gamba rather than by burning fire breaks. This more expensive management option is funded through airport governance processes and customer charges.

In summary, weed governance does not operate in a vacuum. Rather, managers must incorporate it into other layers of governance and regulation including those relating to fire. This adds to workloads and provides little guidance around competing priorities other than when human life and property are at risk, where the urgency of fire risk drives its own priorities. Incorporating weed management into regular land management practice is routinely done where it incurs a minimal cost, where there are opportunity costs or where there is recurrent funding available. Routine, while not a panacea for good management, is certainly a point of traction. However, such traction will not necessarily conform to best practice weed management principles.

Practices of responsibility

Whatever the challenges on individual landholdings, they are magnified across complex tenures, and practices of responsibility are partly spatial. The gamba grass strategic plan acknowledges that weed management is a 'shared responsibility' requiring 'coordination' and 'capacity' (AWC 2013, p. 3). Thus a particular kind of distributed agency is recognised as necessary, but its complexities are often glossed over. Governance proliferates to include a variety of non-state actors (private landholders, business operators, volunteers) acting on a patchwork of property tenures (public land including conservation and other tenures, private land, leasehold, Aboriginal land) with complex layers of responsibility. The particular shapes of different tenure systems are also influential; for example, road verges and powerline easements provide classic corridors for the spread of weeds. Notwithstanding the responsibility of the state to legislate and regulate on issues of invasive plant management, and regardless of how governance mechanisms might attribute legal duty, most study participants agreed that responsibility for land management should be the task of individual land managers. At the same time, participants recognised that the work of weeding (like weeds themselves) must be distributed across and owned by the whole community for it to be effective. Government managers are very aware of this when working with landholders:

> We see it as if you manage land or own land and you have weeds on it then they become your weeds and your problem ... and we try and provide landholders with knowledge and skills to manage them and prevent them from becoming worse or establishing.
>
> (Government Officer C)

As the example in the previous section also exemplifies, government extension programmes are undertaken in the context of trying to 'get people managing their land responsibly' rather than prosecuting or targeting them.

In our interviews we did not encounter any sense of hopelessness or surrender, but instead a sense of individual responsibility for land management was a recurrent theme, with participants referring to the 'independent spirit' of people in the Northern Territory as well as a sense of 'doing things differently'. This may mean for example, finding ways to manage millions of square kilometres of land on one's own. These framings were mostly viewed as a helpful dimension in weed management because of the perception that a sense of self-reliance often translates into individuals 'doing the right thing', even where they are not under any specific obligation to do so. In one case a participant described a sense of community care in his neighbourhood around a Landcare project, and becoming annoyed when it seemed that certain neighbours did not care about their own 'blocks' of land. Similarly this participant described the satisfaction of working on other weeding projects without ever being asked or required to do so. Being required or legally obliged to manage weeds was clearly a point of contention for some, even amongst those who prided themselves on their contribution to weed management. Being legally bound was viewed as yet another bit of government regulation or 'red tape' that had to be dealt with, even though there was also grudging recognition that it might be required where not all land managers were doing the 'right thing'.

The tensions around responsibility and obligation were especially important in the context where government was itself the land manager. Most participants were keenly aware that government had been responsible for introducing gamba grass in previous decades. Some participants were critical of the Northern Territory government's dual role as both land manager and weed authority. During the more private context of our interviews they were quick to point out examples where they perceived government not to be managing weeds appropriately on its own land. As one participant put it, conflating 'government' and 'scientists':

> So, the government's – and the same with the bloody cane toads, which we never had for a long time! Bastard things! So scientists have got a bit to answer for. But now, they can't just say, 'Well, now the weeds are there and you're administering that block of land. It's your responsibility to make sure they're not there.' It's just bloody ridiculous, you know what I mean?
> (Extractive Industry Member)

A different atmosphere prevailed in the more formal stakeholder engagement meetings we observed. We noted that any open critique of changing government perspectives, or of government's performance in managing its own land, created a very uncomfortable atmosphere between government representatives and other land managers. In the context of coming together to deal with contemporary problems, historical 'sticking points' were not aired openly. Tightly scheduled stakeholder meetings provide limited opportunities for more broad-ranging discussion of these issues, thus contributing to continuity rather than circumvention of cynicism and dissatisfaction with the current process.

Policy and governance implications

Returning to the overall questions, we can address the third, most specific question first. What can be learned from these everyday practices of invasive plant management, in which people live with the plants in various ways? Certainly gamba governance will need to be more responsive to the rapidity of ecological change and the ways in which different actors are responding. But it also needs to be alert to the diversity and politics of the practices it recognises and validates. In parallel with the literature on climate change (Bulkeley & Broto 2013), this analysis shows that the governance of invasive species is constituted by diverse, pragmatic practices. Three areas of focus are important.

The practices of priority setting and species declaration show that, contrary to the rhetoric of getting rid of weeds completely, there is already considerable acknowledgement of living with them in policy and in on-ground practice, the most obvious example being the exclusion of species from lists of declared weeds where control is considered unfeasible. At the moment, the cultural-political power of the pastoral industry exceeds both the power of the scientific advice, and affects the social license to enforce the legislation upon which it is based, but in quite hidden ways. (On the other hand, that power was not sufficient to stop gamba being declared.) This also raises questions about the accounting of scientific advice, and how other factors are weighed against such advice. Further, the effectiveness of the state as a credible actor or enforcer is compromised by its history of having introduced and fostered gamba in the first place, and by being seen in some circumstances as a poor manager of its own lands.

The practice of managing the combined gamba grass-bushfire risk assemblage has both advantages and disadvantages. One advantage is that it has mobilised community feelings and galvanised political action, providing a new set of intervention levers through fire policy. The materiality of the plant and fire has rendered risk extremely visible. The gamba example shows that weeds are not only or even best approached through weeds policy but can also be managed through fire policy. Bushfire governance has a longer history than weed governance, stimulated by the threats to people and property. Overall then, there are more systematic coordination processes in place, and the role of volunteers is explicit. Nevertheless evident ambiguity about the responsibilities of landholders and government in fire management (DLRM 2013) indicates that there may be parallel challenges to those of invasive plant management.

One disadvantage of managing weeds through fire policy is that the focus of fire policy on human spaces deflects action from best practice weed management that stipulates a focus on outlier occurrences. Outliers tend to be in the areas of savanna where biodiversity is both high and at risk. So being driven by fire risk diverts effort both from these outlying spaces, and from other invasive species that may be just as problematic, albeit not to humans. Political choices are necessary between competing priorities.

Giving more systematic voice to the messy realities of living with and killing weeds – *practices of responsibility* – will help mobilise a broader range of resources including individual landholders and networks that are coordinated for different motivations. An unlikely example comes out of the pervasive theme in this study of individual autonomy and resistance to government control. (This issue is discussed further in Chapter 9.) In practice, purely legal renderings of land management responsibility will not be enough for effective weed management. Similarly, landholders' perceptions of obligation versus responsible practice do not necessarily map neatly onto one another. The in-depth interviews and participant observation methodology highlight some of the differences between what is said and discussed in formal stakeholder engagement processes, and what landholders are willing to disclose in private about the factors that may inhibit or alter good weed practice. Cynicism directed towards government was often used to justify limited action, but was also a genuine expression of frustration rather than unwillingness to participate. This raises procedural questions, even for more dynamic framings such as adaptive governance, where common goals may not be agreed upon. In a more dynamic understanding of governance, the identification of double standards and weak links in the chain should be valorised as aiding effective practice. Thus cynicism and paralysis could potentially be turned to effective action, but this will require more honest acknowledgement from government of its own historical contingencies, ever-changing government discourses and limited capacity to act.

Paradoxically, we pay less systematic attention to cross-scalar environmental governance issues in the much smaller scale fragments of urban areas. Yet they can be just as important. For example, Parker et al. (2008) showed that the optimum patch and corridor formation to preserve biodiversity in the Shellharbour Local Government Area would need to heavily use both riparian habitat and adjacent residential yard space. Local councils are only just beginning to consider how to utilise neighbours and neighbourliness in the conservation of biodiversity. As we see in Chapter 9, this is where expressions of autonomy and individual freedom are also very strongly expressed.

What do these findings mean for the second question; how well suited is the current policy process to more mobile and distributed agencies? The recognition of invasive plants as a global environmental issue, and one likely to become more significant in particular regions under climate change, has led to important steps in cross-scalar and cross-jurisdictional coordination, such as the Australian Weeds of National Significance programme. It is clear from the gamba study how much governance work takes place in the spaces between policy and action. The ongoing tension here is how a more responsive and open process of governance incorporates learning and deals with failure. With enormous biodiversity loss at stake, the rationale of invasive plant governance should no longer be framed around absolute control, which is impossible, in order to clarify some of the political and policy choices involved in living with invasives. Because this issue is so important, the most effective governance practices need to be identified. Weeds exceed and escape governance in multiple ways, and many aspects of on-ground management practice already reflect much more nimble, pragmatic engagements than the top-down policies imply.

To return then to the broader implications of our first question; how should we act when the total context may be ungovernable? This is not only a question for some future time, called the Anthropocene. A focus on the practices of governance shows the many ways that the ideal of present control is illusory. But there are also many ways in which vernacular experience and knowledge provides effective accommodations with invasive plants. This particular study provides one illustration of the relational rather than separationist interventions discussed in Chapter 4. It is here that the 'lessons' for other issues such as climate change must be found. There are no fixes, and there can be no fixity, but there can be living – and dying – together.

Notes

1 An earlier version of this chapter was published as Head and Atchison (2015). More details and fuller referencing can be found in that paper.
2 www.environment.gov.au/biodiversity/invasive/weeds/weeds/lists/wons.html. Last accessed 16 August 2015.

References

Anon. 1987. "Gamba grass." *Tropical Grasslands* 21(1): 44–46.

Atchison, J. 2015. "Experiments in co-existence: The science and practices of biocontrol in invasive plant management." *Environment and Planning A* 47. DOI: 10.1177/0308518 X15597106.

AWC (Australian Weeds Committee). 2013. *Weeds of National Significance Gamba Grass (Andropogon gayanus) Strategic Plan 2012–2017*. Canberra: Australian Weeds Committee.

Barker, K. 2014. "Biosecurity: Securing circulations from the microbe to the macrocosm." *The Geographical Journal*. DOI: 10.1111/geoj.12097.

Beeton, R. n.d. *Advice to the Minister for the Environment, Water, Heritage and the Arts from the Threatened Species Scientific Committee (the Committee) on Amendments to the List of Key Threatening Processes under the Environment Protection and Biodiversity Conservation Act 1999 (EPBC Act)*. Accessed 26 June 2014. Available at www.environment.gov.au/node/14590.

Brennan, B. 2011. "Catastrophic fire danger warning for Top End." *ABC News*, 23 August 2011. Accessed 11 August 2015. Available at www.abc.net.au/news/2011-08-23/20110823catastrophic-fire-alert/2851504.

Bulkeley, H. and V. C. Broto. 2013. "Government by experiment? Global cities and the governing of climate change." *Transactions of the Institute of British Geographers* 38(3): 361–375.

Bulkeley, H., M. Watson and R. Hudson. 2007. "Modes of governing municipal waste." *Environment and Planning A* 39(11): 2733–2753.

Cook, G. D. and L. Dias. 2006. "Turner Review No. 12. It was no accident: Deliberate plant introductions by Australian government agencies during the 20th century." *Australian Journal of Botany* 54(7): 601–625.

DLRM (Department of Land Resource Management). 2013. *Discussion Paper: Review of the Bushfires Act 2009*. Winnellie, NT: Department of Land Resource Management.

DLRM (Department of Land Resource Management). 2014. "New boundaries to manage Gamba grass." Media note, 9 April 2014, Palmerston, NT: Department of Land Resource Management. Accessed 1 July 2014. Available at www.lrm.nt.gov.au/news-room/new-boundaries-to-manage-gamba-grass#.U7I1w6N--Uk.

Driscoll, D. A., A. Felton, P. Gibbons, A. M. Felton, N. T. Munro and D. B. Lindenmayer. 2012. "Priorities in policy and management when existing biodiversity stressors interact with climate-change." *Climatic Change* 111: 533–557.

Duff, N. and J. K. Weir. 2013. *Weeds and Native Title – Law and Assumption*. Canberra: RIRDC Publication No. 13/078.

Duit, A. and V. Galaz. 2008. "Governance and complexity – emerging issues for governance theory." *Governance: An International Journal of Policy, Administration and Institutions* 21(3): 311–335.

Enticott, G. 2014. "Relational distance, neoliberalism and the regulation of animal health." *Geoforum* 52: 42–50.

Epanchin-Niell, R. S., M. B. Hufford, C. E. Aslan, J. P. Sexton, J. D. Port and T. M. Waring. 2010. "Controlling invasive species in complex social landscapes." *Frontiers in Ecology and the Environment* 8(4): 210–216.

Flores, T. A., S. A. Setterfield and M. M. Douglas. 2005. "Seedling recruitment of the exotic grass *Andropogon gayanus* (Poaceae) in northern Australia." *Australian Journal of Botany* 53: 243–249.

Franklin, D. C., A. M. Petty, G. J. Williamson, B. W. Brook and D. M. J. S. Bowman. 2008. "Monitoring contrasting land management in the savanna landscapes of northern Australia." *Environmental Management* 41(4): 501–515.

Head, L. and J. Atchison. 2015. "Governing invasive plants: Policy and practice in managing the Gamba grass- (*Andropogon gayanus*) bushfire nexus in northern Australia." *Land Use Policy* 47: 225–234.

Higgins, V. 2004. "Government as a failing operation: Regulating administrative conduct 'at a distance' in Australia." *Sociology* 38(3): 457–476.

Law, J. 2006. "Disaster in agriculture: Or foot and mouth mobilities." *Environment and Planning A* 38(2): 227–239.

Lockie, S. 2004. "Collective agency, non-human causality and environmental social movements: A case study of the Australian 'landcare movement'." *Journal of Sociology* 40(1): 41–58.

Lockie, S. and V. Higgins. 2007. "Roll-out neoliberalism and hybrid practices of regulation in Australian agri-environmental governance." *Journal of Rural Studies* 23(1): 1–11.

Mann, C. and J. D. Absher. 2014. "Adjusting policy to institutional, cultural and biophysical context conditions: The case of conservation banking in California." *Land Use Policy* 36: 73–82.

March, N. 2011. *Weed Management Guide, Weed of National Significance Gamba grass (Andropogon gayanus)*. Accessed 2 July 2014. Available at www.weeds.org.au/WoNS/gambagrass/docs/47053_ERGO_Weed_Mgmt_guide_GAMBA_GRASS_spread.pdf.

McGeoch, M. A., S. M. H. Butchart, D. Spear, E. Marais, E. J. Kleynhans, A. Symes, J. Chanson and M. Hoffmann. 2010. "Global indicators of biological invasion: Species numbers, biodiversity impact and policy responses." *Diversity and Distributions* 16(1): 95–108.

NRMMC (Natural Resource Management Ministerial Council). 2007. *Australian Weeds Strategy—A National Strategy for Weed Management in Australia.* Canberra: Australian Government Department of the Environment and Water Resources.

Ogden, L., N. Heynen, U. Oslender, P. West, K.-A. Kassam and P. Robbins. 2013. "Global assemblages, resilience, and Earth Stewardship in the Anthropocene." *Frontiers in Ecology and the Environment* 11(7): 341–347.

Parker, K., L. Head, L. A. Chisholm and N. Feneley. 2008. "A conceptual model of ecological connectivity in the Shellharbour Local Government Area, New South Wales, Australia." *Landscape and Urban Planning* 86(1): 47–59.

Preece, N., K. Harvey, C. Hempel and J. C. Z. Woinarski. 2010. "Uneven distribution of weeds along extensive transects in Australia's Northern Territory points to management solutions." *Ecological Management and Restoration* 11(2): 127–134.

Rossiter, N. A., S. A. Setterfield, M. M. Douglas and L. B. Hutley. 2003. "Testing the grass-fire cycle: Alien grass invasion in the tropical savannas of northern Australia." *Diversity and Distributions* 9(3): 169–176.

Rossiter-Rachor, N. A., S. A. Setterfield, M. M. Douglas, L. B. Hutley, G. D. Cook and S. Schmidt. 2009. "Invasive *Adropogon gayanus* (Gamba grass) is an ecosystem transformer of nitrogen relations in Australia's tropical savanna." *Ecological Applications* 19: 1546–1560.

Settele, J., R. Scholes, R. Betts, S. Bunn, P. Leadley, D. Nepstad, J. T. Overpeck and M. A. Taboada. 2014. "Terrestrial and inland water systems." In *Climate Change 2014: Impacts, Adaptation, and Vulnerability. Part A: Global and Sectoral Aspects. Contribution of Working Group II to the Fifth Assessment Report of the Intergovernmental Panel on Climate Change*, edited by C. B. Field, V. R. Barros, D. J. Dokken, K. J. Mach,

M. D. Mastrandrea, T. E. Bilir, M. Chatterjee, K. L. Ebi, Y. O. Estrada, R. C. Genova, B. Girma, E. S. Kissel, A.N. Levy, S. MacCracken, P. R. Mastrandrea and L. L.White, 271–359. Cambridge: Cambridge University Press.

Setterfield, S. A., N. A. Rossiter-Rachor, M. M. Douglas, L. Wainger, A. M. Petty, P. Barrow, I. J. Shepherd and K. B. Ferdinands. 2013. "Adding fuel to the fire: The impacts of non-native grass invasion on fire management at a regional scale." *PLoS ONE* 8(5): e59144.

Simberloff, D., I. M. Parker and P. N. Windle. 2005. "Introduced species policy, management and future research needs." *Frontiers in Ecology and Environment* 3(1): 12–20.

Tassin, J. and C. A. Kull. 2015. "Facing the broader dimensions of biological invasions." *Land Use Policy* 42: 165–169.

9 Beyond fortress and sprawl

Retrofitting cities, suburbs and households

Suburbs are villains in much environmental thinking. They sprawl across good agricultural land and green belts rich in biodiversity, covering porous natural surfaces with concrete and other impermeable materials. They stretch infrastructure so thinly that it is more expensive than necessary (sewerage) and cannot possibly be effective (public transport). Well planned city centres, on the other hand, are seen as the lifeblood of civilisation; places where humans can live at high densities with all the creative flowering that comes as a result; leaving nature to flourish somewhere else, away from the destructive hand of humans, who may visit but will probably not stay.

In practice it is not that simple. In Australia, suburban nation par excellence, research shows how many people value suburban life for the combination of privacy, freedom and connection to nature that it affords. We know that the separate house and garden continues to be the favoured residential form for most Australians (Gleeson 2006, Timms 2006). Suburbs can be centres of creativity as much as urban centres (Brennan-Horley & Gibson 2009). Recent shifts in densification of urban areas, with higher proportions of households living in apartments or units, are driven partly by shifts in preferences, but also by the spiralling cost of housing relative to average wages. Even in the densest parts of a city such as Sydney, the topography of the landscape means that harbour, river or creek landscapes, with their combination of green and blue, are never far away.

In the spirit of finding hope in unlikely places, this chapter focuses on the rich vein of life and living in suburban households. Suburbs need to be part of our Anthropocene thinking for two reasons. First, there is not time to completely remake our cities before climate change impacts disrupt their functioning. Retrofitting rather than starting again is a crucial strategy. We should think of retrofitting not just our buildings, but also our everyday practices, work patterns and use of time. Second, the fortress city model of high-density urban living is totally dependent on fossil fuel driven energy flows from outside the city. Through the metaphor of the ecological footprint, we are in a better position to understand how deeply cities depend on the surrounding and far-flung places that supply their underpinning agricultural assets. In a decarbonising economy (whether a policy choice or forced on us by system collapse), suburbs must be viewed rather differently. They provide a range of opportunities for water capture and storage, energy generation, food production and social support. They can be thought of as zones of production rather than (just) consumption (Gibson-Graham 2011).

This chapter uses the suburb as a conceptual and material liminal zone between urban and rural, and between other boundaries (Head & Muir 2006). The messiness of suburbs provides a resource for reflection, in the same way that cities are likely to become (and in practice already are) much more messy and mixed. I do not abandon the 'city' here, but I want to draw it into a continuum or a set of relations with suburbs, agriculture and distant places. These connections have been encapsulated by Whitehead: 'within the Anthropocene it is important to recognize the lines of geographical connection that join supermarkets in the UK with peat swamps in Indonesia, and the cooking and eating practises favoured in the US kitchen with global climatic change' (2014, p. 9).

Although such relations will be subject to various kinds of disruption, it is unlikely they will be severed altogether. The arguments in this chapter are also informed by work in urban political ecology that renders a clear distinction between the city and nature meaningless (Gandy 2002, Heynen et al. 2006, Houston & Ruming 2014). Cities are understood as relational entities, underpinned by production in faraway places.

This chapter also uses the household as a scale of analysis. Households are social assemblages with variable gender, age, class, ethnic and familial structures. They contain their own politics and practices, and interact with wider social structures in complex ways. An important focus for climate change policy is households of the developed world, which contribute to climate change (Reid et al. 2010) through greenhouse gas emissions both from direct energy use and as conduits for the flow of goods and services (Druckman & Jackson 2009). Households make sense both to the people who live in them, and to government policy-makers, as foundational social units, and as sites through which it is logical to understand the consumption of energy, water and materials that have implications for sustainability issues and climate change mitigation. They are also an important potential scale of climate change adaptation, although climate change adaptation research in relation to the developed world is only just beginning to systematically engage with the household scale (Keogh et al. 2011, Sherrieb et al. 2010, Toole et al. in prep.). Everyday life in such households is at once insignificantly tiny and the engine of the global economy that drives climate change. This will be increasingly the case with growing levels of affluence in countries such as China (Peters et al. 2007, Zheng et al. 2011) and India (Kadian et al. 2007). Cities and households are both scales at which the practice of hope can feel more tractable – we can imagine what everyday life would look like.

The material in this chapter should be viewed through two different lenses. The first is an environmental sustainability lens. What is the scope to improve our practices, and to integrate behavioural, infrastructure and systemic change? In understanding households, suburbs and cities as complex assemblages or networks, it is necessary to think of how any particular intervention will generate traction or friction in the wider system (Head, Farbotko et al. 2013). This is a difficult challenge given how deeply fossil fuel dependence is embedded in all aspects of our lives, from transport infrastructure to food production to patterns of work. The household is an essential site to analyse the dilemmas and potential of 'scaling up' climate solutions from cultural analysis of the complexities of everyday life (Liverman 2008, Abbott & Wilson 2014). It follows from the urgency of the challenges that behavioural change must be understood in a very particular light:

> the greatest behavioural leverage comes from those insights that help people act and prevent fossil fuels from being extracted, while the current behavioural emphasis on energy efficiency gains, though necessary and important will be insufficient unless we keep this core objective firmly in mind.
> (Rowson 2013, p. 5)

The second lens – a more catastrophic survival lens – is more difficult, but it does throw up new opportunities and inverts some of the understanding of where vulnerability and capacity can be found (Gibson et al. 2015). How much food could I grow in my garden if I had to? What if I could not buy any more towels, or nails, or sewing thread for the rest of my life? What if the electricity was only on for two hours a day and petrol was rationed? What if two refugees had to live in my house? What sorts of caring and sharing propensities and possibilities are there in my street? For those of us in the rich, twenty-first century world, such questions tend to invoke the past – stories from parents and grandparents who grew up in Depression Era and wartime austerity. Our best example of collective sacrifice is usually presented as World War II Britain, with its campaigns to Dig for Victory and Make Do and Mend. Indeed, such campaigns are often invoked in the cause of contemporary environmental sustainability, albeit in conservative ways 'that pose austerity and self-sufficiency as a solution to economic inequalities' (Ginn 2012, p. 295). Yet we do not need to look to the distant past to find catastrophic circumstance or dormant capacity; there is plenty of each in today's world. As Howitt et al. argue in their discussion of the colonial framing of emergency management, in many indigenous communities 'everyday life proceeds in a constant state of emergency because of the historical context in which people find themselves' (2012, p. 55).

The rest of the chapter is structured as follows. I first examine some aspects of suburban form, as an example of how dormant resources can be identified and considered. Then I discuss households as places where vernacular capacities take particular expression, albeit the capacity for transformative change is constrained in a number of ways. Strong evidence for adaptiveness provides hope that capacity is broader and deeper than we currently imagine. Connecting to themes of earlier chapters, this discussion provides a way to imagine humans as forces for environmental good, not just damage. Three particular themes in this work are then pursued in greater depth to provide examples of practical action. It is not necessary to develop 'green' identities to foster strong environmental outcomes. It is necessary to acknowledge the pervasiveness of values around family, autonomy and privacy. Rethinking abundance and scarcity does not mean

Identifying resources in suburban form

The quest for better patterns of development has polarised debate around two main urban forms: compact and sprawl. Compact developments can greatly reduce transport fuel use and related greenhouse gas emissions by minimising car dependence in cities (Newman & Kenworthy 1999). The greater capacity of high-density environments to provide sustainable transport benefits and to protect ecologically sensitive land areas and water quality than lower-density developments (EPA 2006) has initiated a wide acceptance of urban intensification as a sustainable solution. However, questions have been raised about the environmental and ecological impacts of urban consolidation as decreasing garden sizes and tree canopy cover could result in loss of biodiversity and increased surface sealing (Pauleit et al. 2005). Imperviousness has critical impacts in reducing ground water recharge and evapotranspiration, increasing storm water run-off quantity and declining storm water run-off quality (Hasse & Nuissl 2007).

The link between the household scale and wider urban morphologies has been the subject of a range of recent research. Gardens are a dominant land-use type and in total could occupy up to one-third of an urban land area (Mathieu et al. 2007; Gaston et al. 2005). Incremental changes over time in house sizes, residential imperviousness and garden areas in neighbourhoods could alter urban morphologies (Perry & Nawaz 2008). Urban tree canopy cover can protect environmental quality and human health through carbon sequestration and storage and air quality improvement (Nowak & Crane 2002). The Biodiversity in Urban Gardens in Sheffield (BUGS) project in five cities of the UK has shown that larger gardens can support a variety of land-cover types that are important for achieving improved ecological and environmental functions (Smith et al. 2005; Loram et al. 2008). In the Australian context, we have already discussed how private gardens are important links in reconnecting biodiversity corridors (Parker et al. 2008).

As we consider how extensive urban swathes must adapt to climate change, it will be necessary to understand the cumulative sustainability potential of millions of private homes and their associated outdoor spaces. It will be important and beneficial to retrofit gardens as well as houses, although the latter have received more attention in public discussion. It is perfectly conceivable that future land capability analyses will consider not only the broad sweep of peri-urban zoning and agricultural lands, but will seek to identify micro-spaces within cities and suburbs that can be utilised for a range of social goods, including the production of energy, water and food. Detailed mapping from remote sensing and GIS (geographic information system) now makes this possible, and informal local groups are already doing this through, for example, mapping wild food availability in urban areas. Gardens have received attention in relation to water production and consumption, via water tanks, and in relation to food production. In one example, Gibson-Graham discusses the productive role of gardens in response to Hayden's Field Guide to Sprawl:

> It is easy to see this landscape as an inert, durable, obstruction to living differently. But our expanded regional profile might act as an antidote field guide, one that begins to identify emergent habitats, develop new lexicons and get a worm's eye view of possible landscapes of sustainable growth. From this inquiry we might discover vibrant matter we can work with, new dynamics of change that we can move with and against. We might work with people in place to push at elements of the assemblages we are part of to enact new modes of living.
>
> (2011, pp. 15–16)

The speculative diagram reconfigures an aerial photo of sprawl:

> to show networks of economic reciprocity and non-monetized exchange, distributed systems of food, energy and water production and habitat restoration for frogs and other animals. These pieced-together, imagined connections are potentially the emerging habitats in which a new mode of humanity might be nurtured into life.
>
> (2011, p. 16)

Sumita Ghosh and I (Ghosh & Head 2009) did some close analysis, comparing the morphologies and sustainability potential of two different types of suburbs. We knew that research had shown residential backyards to be potentially important sites of waste composting, local food production, native vegetation restoration, biodiversity promotion and promotion of sustainable water consumption practices. Considering potential residential occupancy and roof areas of different house typologies, a significant proportion of domestic water requirements can be collected from residential roofs (Vale & Ghosh 2006). There is considerable debate over the role of home-based food production in contributing to food security and achieving sustainability at a neighbourhood scale. However, limited research had been conducted on the structures of cumulative residential garden spaces in conventional and new suburbs and their corresponding sustainability potential.

Our sample encompassed dwellings we referred to as 'Traditional' (larger blocks, small houses, typically built in the 1950s and 1960s) and 'Modern' (smaller blocks, large houses, typically built from the 1990s on). In the Modern, 98 per cent of the block sizes are between 400 and 650 m^2, while 59 per cent of the garden areas in the Traditional are in this range. We focused on the outdoor spaces – roofs as sources of solar energy and water capture, and gardens for a range of environmental goods (for details of methods and results, see Ghosh & Head 2009). Land-cover patterns showed that the percentage of permeable cover in the Traditional was nearly double that in the Modern, while the average building roof size in the Modern is nearly twice that of the Traditional. Though front garden areas were similar in the two case studies, the rear garden areas with trees were larger in the Traditional, forming a green open space corridor. In the Modern, this open space connectivity was frequently interrupted by features such as paving or swimming pools.

Our analysis showed that Traditional development with smaller roof sizes could supply slightly more than two-thirds of their total water demand. Modern development with larger building roof sizes had the potential to harvest the total water demand of the households, plus an additional 35 per cent, provided appropriate tank infrastructure was installed. The Hills Hoist clothes line (and its successors) for drying clothes outdoors is often touted as an important Australian invention. Both case study suburban forms had plenty of space for this activity, which effectively contributes to energy savings and CO_2 emissions reduction.

Traditional development has a greater capability for supporting ecological functions because of the availability of connected green spaces, ample tree canopy cover and more permeable and productive land areas than the Modern. The Traditional development with nearly four times higher tree canopy cover than the Modern allows significantly higher per capita carbon storage and sequestration. Separating tree canopy cover land areas in the productive land area calculation allowed us to account for those land areas which could be practically engaged in local food or vegetable production. The food production potential of the Traditional development is high because of greater availability of productive land areas.[1]

Nevertheless, cultural research into gardens also alerts us to the dangers of focusing on physical form while ignoring the social and ecological relations that constitute and interact with that form. It is to this theme that we now turn.

Capacity and constraint in the connected household

In affluent urban societies households are an increasing focus of government policy in relation to sustainability issues, and an expanding research literature considers the household as a crucial scale of social organisation for pro-environmental behaviour (Reid et al. 2010, Gibson et al. 2013, Lane & Gorman-Murray 2011, Tudor et al. 2011). There is considerable potential for reducing emissions via behavioural changes such as installing low-flow showerheads, changing driving behaviours and line-drying of laundry (Dietz et al. 2009), but there is also considerable research showing why behavioural change is not straightforward (Lorenzoni et al. 2007, Ockwell et al. 2009, Shove 2010). For example, smart meters do not challenge practices that householders consider non-negotiable (Strengers 2011). Most incentive and education programmes pay little attention to the ways household energy, water and other resource consumption practices are part of the rituals, rhythms, habits and routines of everyday life (Shove 2003, Gregson et al. 2007). Sustainability campaigns normally fail to appeal to, or appreciate, the emotional meanings attached to material possessions (Hobson 2008) or home spaces (Blunt & Dowling 2006). Even when householders want to make behavioural changes there are a range of ways they can be locked in.

Our collaborative research has been part of this broader endeavour over the last few years. We framed households as part of, and the product of, a network of connections.[2] In the context of this book I want to draw out two particular threads that relate to the practice of hope; a broader understanding of vernacular capacities, and the potential to generate transformative change. To what extent can a range of suburban experiments – not all of them with green labels – be generative? Or, under worst-case scenarios, where are the survival capacities and risks? Here I draw on an analysis in which we sought a means to scale up our household cultural research and develop systematic comparisons with other Australian work of similar scope. It was important to keep the richness and detail, and not lose the complexities and contradictions in this kind of data. We used a technique called meta-ethnography.[3] Notwithstanding considerable debate over the methodology, its key contribution is that it 'does not conceptually dismiss single case studies as locally bound', but rather 'compels us to acknowledge the importance of not only the uniqueness of individual cases, but also the uniqueness of collectives' (Doyle 2003, p. 340). Seven persistent and inter-connected themes, summarised in Table 9.1, were identified.

Table 9.1 Themes identified in meta-ethnography of household sustainability analyses

Theme	Summary
Family is central	Family and social relationships are key drivers of household decision-making, even in environmentally conscious households. Roles, obligations and practices of care strongly influence environmental outcomes.
Adaptiveness is contingent, but pervasive	Everyday life requires and enhances many types of adaptiveness, flexibility and coping. Households have differential capacities and potential for adaptiveness; their starting point is to work with what they have and where they are.
A sense of *autonomy* needs to be maintained	Power is constantly negotiated at a range of scales (individual, family, household). Freedom, choice and control are articulated as important issues – in ways that variously conform with and confound wider governmental or policy objectives.
Households make sense of the world through *materials*	Everyday interaction with and use of material things, including the materials of the home itself, are at the forefront of people's awareness. In interviews they constantly bring abstract concepts, including climate change, back to the 'stuff' around them.
Boundaries are dynamic and subjective	The family and home provide privacy, freedom and retreat from the world, but are variably permeable to friends, visitors and networks of sharing.
Time is an important currency	Everyday life involves constant attempts to 'smooth' and 'save' time to make routines seamless and minimise interruptions. In practice these rhythms and movements are subject to disruption and friction, especially where children are involved.
Paradoxes abound	All the themes are cross-cut by paradoxes that hold practices, values and perceptions in tension. Consistent with other research, these findings demonstrate that identities and behaviours do not align.

Source: Head et al. (in prep.).

The most positive implication in this analysis is the strength of adaptiveness. It is clear that households have all kinds of capacities to respond, cope and adapt – many of which may not be readily apparent when assessing resilience at either the individual or the population scale (Downes et al. 2013). They may be readier to make sacrifices to deal with climate change than governments and policymakers have given them credit for. Income, education, social class and remoteness have become common proxies for vulnerability, but neither vulnerability nor capacity should be assumed from macro-scale demographic or socioeconomic data (McNamara & Prasad 2014). Developed world populations determined as vulnerable using quantitative demographic data are being shown through subsequent qualitative methods to have strong social bonds, from prior experiences of rallying together in response to extreme external forces such as droughts, wildfires and floods (Beer et al. 2013).

Further, some inversions of capacity and vulnerability are likely (Toole et al. in prep.). Many indigenous communities with macro-scale vulnerability indicators also have strong kinship networks and capacities to live off the land (Howitt et al. 2012). Refugees who were evacuated due to flash flooding in the Queensland floods of 2011 were able to cope well due to their prior experiences of displacement. In this instance, being twice displaced was a source of strength rather than vulnerability; many stated that their experiences of 'very difficult times back home' had helped them to cope with the flood disaster. They had prior experience of evacuating and of how to organise and support neighbours in the aftermath of upheaval, and were able to put this into practice (Correa-Velez et al. 2015). Adams contrasts attitudes to uncertainty and change in different cultural contexts:

> While indigenous and local communities have particular cultural characteristics adapted to conditions of risk and uncertainty, modern and modernizing societies have quite different cultural characteristics that might make them particularly vulnerable to rapid and unwanted change. These societies attempt to control change, to maintain stability, to impose a form of order that facilitates predictable outcomes. But these norms are surprisingly recent: only a couple of generations ago in developed world contexts, frugality, stoicism, preparedness for hardship were not only normal attitudes but celebrated as strengths.
>
> (2015, pp. 21–22)

Clearly there is much to learn from communities with diverse experiences. Of course, there is the risk of valorising poverty in this debate, and it is important to recognise the profoundly unequal distribution of resources. Mee et al. remind us of this in relation to the 'resources for adaptation', particularly for renters: 'the limited access of many tenants to "resources for adaptation" such as gardens, water efficiency, and alternative energy . . . is exacerbated by regulatory practices, including leases, insurance, and capital investment, that help shape the socio-natural relations of tenure' (2014, p. 365). Nevertheless, as more catastrophic climate scenarios increase in likelihood, these discussions are necessary.

Challenge to green subjectivities

One clear implication from the exercise of identifying or cataloguing cultural resources not understood as 'green' is that it is not necessary to identify as an environmentalist to be making important contributions to environmental sustainability. It is quite clear now from a range of research that practice, identity and attitudes do not necessarily line up. But the value-action 'gap' (Blake 1999) can also be reconfigured as a value-action 'opportunity' (Hitchings et al. 2015). A lot of the work of sustainability is not done for environmental reasons, but for reasons of frugality, or because people dislike waste. Some of the most avid water savers express vehemently anti-green attitudes (Sofoulis 2005), drawing instead on a rhetoric and identity of frugality; and a lot of sustainability work is being done by low-income households who do not necessarily identify as 'green' but who nonetheless consume less (Waitt et al. 2012). As mentioned above, this is particularly the case among older people and retirees. These examples provide great potential to support the maintenance and retrieval of past practices, and to catalogue and share knowledge not yet lost. They suggest that policy and governance should focus on the frictions in infrastructure and practice, rather than trying to foster green subjectivities. On the other hand, a broader range of subjectivities and identities provide cultural resources with which to work, such as appealing to ideas around thrift, sufficiency and not being wasteful. Particularly important is the constellation of ideas around the family, privacy and freedom.

Privacy, autonomy, family

The finding that the concept of family is central at a household scale of social life is surely a 'no-brainer'. But there is more than meets the eye to the idea that even in strongly green identifying households, the social bonds trump the environmental ones. 'Family' provides a promising non-environmental lever that could be mobilised in climate change response. For example, shared valuing of the family – whatever its diverse forms – provides a potential bridge across the current left-right divide on climate change policy in Australia, the USA and elsewhere (McCright & Dunlap 2011). It is consistent with other cultural environmental research showing the importance of privacy and freedom in the context of domestic life (Blunt & Dowling 2006).

The findings strengthen the basis on which comparisons between households can start to be made at an international scale, particularly from parts of the world acknowledged to be facing similar challenges to Australia, such as the USA (Dietz et al. 2009), and/or where there is a strong tradition of household-scale sustainability research, such as the UK (Tudor et al. 2011). The paradoxes and contradictions have been widely reported in other literature, for example, confirming that a focus on encouraging green identities is a likely barrier and may be counterproductive, while the above-mentioned family values – albeit in diverse expressions – may provide a non-environmental lever that can gain wide traction. In contrast, the focus on autonomy and privacy may be a more uniquely Australian concern.

Rethinking materiality, abundance and scarcity

There are several ways in which we need to think more carefully about materials and material relations. Certainly most of us could vastly simplify the amount of 'stuff' in our lives without detriment to our wellbeing. However there are several reasons why we should not confuse a reduction in consumerism with an anti-material or amaterial view of the world. As Jackson argues, the appealing idea that we might do away with material things once basic material needs are satisfied: 'flounders on a simple but powerful fact: material goods provide a vital language through which we communicate with each other about the things that really matter: family, identity, friendship, community, purpose in life' (2009, p. 143). Indeed it is impossible in any case to disentangle ourselves from the material relations that make up our world. Even those aspects of life that we consider virtual or immaterial depend at base on material relations. The internet that underpins virtuality is material through the supply of electricity and through the complex assemblage of common metals and rare earths that constitute our screens and devices. Air travel may claim to meet experiential rather than physical needs, but its provision is one of the most damaging contributors to the materiality of the atmosphere. (In this respect the characterisation of values as 'postmaterialist' (Inglehart 1997) is quite misleading.) The assumption that a shift from industrial to service domination of the economy automatically frees us from material responsibility ignores these underpinnings. Carr and Gibson (2015) do important work recuperating the material in geography, beyond false binaries of manufacturing and craft production. Among other things they draw attention to the knowledge and skills that exist in the heart of the industrial complex, such as the capacities of steelworkers to creatively repair and reuse materials.

The centrality of engagements with physical things, resources and materials in empirical studies shows that encouraging reduced consumption as a climate change response should not be presented as an attempt to dematerialise everyday life. In fact these kinds of efforts might lead to more active resistance. Rather, new kinds of relationships with things will need to be fostered, building on existing capacities such as the fact that people do not like throwing away things that work (Gibson et al. 2013), or that extended families have well-established processes for storing, sharing and passing on furniture, clothes and other household goods (Klocker et al. 2012). As before, these new kinds of relationships build on, and celebrate, capacities that are already in existence. Examples include aesthetics that celebrate reuse, respect for the embodied energy and labour in physical things and sharing the skills needed for repair (Carr and Gibson 2015).

Discussion of materials and materiality connects to the questions of abundance and scarcity that we discussed in Chapter 6. Consider the example of water. The 'Millennium Drought' in southeastern Australia in the first decade of the century spawned many changes in urban domestic water practices. Dam levels were reported on TV weather forecasts along with the Southern Oscillation Index. Householders cut water consumption in homes and gardens under the influence of water restrictions, installed water tanks and debated recycling and desalination. In Melbourne people grieved the loss of majestic European street and park trees that had survived for a hundred years. In our studies (Head & Muir 2007), households demonstrated unexpectedly high degrees of willingness to endure and used creative practices to capture, save and reuse water outside formal regulation. This was encapsulated by the metaphor of the 'bucket in the shower', that was then used to water the garden. In another example, the climate change scepticism of retired couple Tom and Joan sat in stark contrast with their enthusiasm for water-saving practices, including coordinating their toilet visits to minimise flushing, and capturing grey water from the shower. Our research, mostly in Sydney and Wollongong, chimed with findings from other projects in Brisbane and Melbourne. These were not isolated findings. Why can't we always do this, we asked?

But what happened when it started raining, indeed flooding, in 2010–2011? Reduced water use practices did not appear to lock in permanently from this experience of drought (as the incrementalist behaviour change view might hope). And even during the drought, practices of frugality coexisted with what we called strong watery desires. Should we see this as a sustainability failure? It seems that households are prepared to 'go with the flow' and enjoy or even indulge in abundant water in times of excess. The latter seemingly 'unsustainable' practice viewed through an alternative framing shows a stronger sense of connectedness with the resource's materiality, and with cycles of abundance and scarcity.

The themes of materiality, abundance and scarcity are drawn together by Strengers and Maller (2012) in their study of approaches to water and energy use in migrant households. Undertaking interviews with several generations, they identified three different eras.

Era 1, characterised by *materiality, scarcity* and *diversity*, refers to the 1950s–1970s in the migrants' countries of origin. This era bundles together arrangements 'characterised by regular intermittency and disruption of centrally managed systems, minimal treatment and disposal infrastructure, scarcity, and elaborate collection techniques for multiple energy and water sources' (2012, p. 757).

Era 2, characterised by *immateriality, abundance* and *homogeneity*, refers to the 1970s–1990s in Australia. Migrants accessed 'abundant, cheap and available water and energy sources', in which 'the creation, storage and delivery of resources were largely hidden or obscured from view, rendering them immaterial in householders' everyday lives' (2012, p. 757).

Era 3, the 1990s–2010s in Australia, sees the re-emergence of *materiality, scarcity* and *diversity*. During this period influences such as drought, peak electricity demand and climate change 'partially repositioned water and energy as scarce and valuable resources. These resources have gained visibility in everyday life as householders attempt to save, collect, store and reuse them in inventive ways' (2012, p. 757).

They argue that 'energy and water systems that are materially present, exhibit traits of scarcity, and encourage diversity through innovation, may engage householders as co-managers of their everyday practices' (2012, p. 760). In this analysis, water is much more materially visible and present in its flows through the household than energy, which still has characteristics of Era 2. That may be shifting somewhat with the widespread adoption of solar and hot water PV (photovoltaic) systems; Strengers and Maller note 'that policy making to support the growth of distributed systems has followed rather than led the groundswell of interest' (2012, p. 761). They conclude that:

> policy makers may inadvertently reduce householders' capacity to respond and adapt to climate change impacts by prioritising the resource characteristics of immateriality, abundance and homogeneity. We conclude that policy which prioritises the resource characteristics of materiality, diversity and scarcity is an important, underutilised and currently unacknowledged source of adaptive capacity.
>
> (2012, p. 754)

A key gap here though is abundance, and how we should conceptualise and engage with it. Part of our energy problem is that we have more than we can safely use, through the long-term development of cheap coal. A narrative of scarcity will not be the answer to keeping coal in the ground.

A perhaps unlikely example is provided by Christmas, the West's archetypal expression of abundance and excess – in presents, food, commercialism, hospitality, family intensity. Carol Farbotko and I (Farbotko & Head 2013) explored the role of environmental considerations in Christmas gifting practices among a group of environmentally attuned households in the lead up to Christmas 2010. Our overall finding was that the identity of the green consumer can operate very differently in the context of gift-exchange than in the context of non-gifting consumption; sensibilities towards scarcity and restraint that might operate throughout the year operate differently around Christmas gifting. As one example, Jean's family of four had what we classified as very high green identity, sharing bath water each night that they then reused for toilet flushing. The only cleaning products used in their home were bicarbonate of soda and vinegar. Yet Jean's apparently deep sense of environmental care did not appear prominently in her Christmas, which was a fairly straightforward negotiation of balancing financial concerns with a celebration of family life. Commercially acquired items were an important feature of their Christmas stocking traditions – even though her children were now aged between 16 and 24. Jean talked of going to get stationery and stocking fillers and 'all sorts of junk'. Presents were all wrapped individually – a triple pack of underwear was undone and rewrapped separately to put under the tree, because 'it's opening the presents' that is important. For Jean, everyday frugalities such as saving bath water were not to be carried over into Christmas. Indeed her Christmas practices express a conflicting social concern with maintaining a particular level of financial and material abundance at Christmas, even in the face of financial hardship.

Just as non-green identities provide examples of unheralded capacities, this contradiction of apparently green identity alerts us to the risks and intransigence around abundance. We do abundance well when it expresses celebration, sharing and hospitality. Or when 'stuff' can be transformed into other lives – shower water can be recycled for the garden, and unwanted Christmas presents can be given to the second-hand shop (after a suitable period in the cupboard to render their gifting properties inert). But there is little evidence in any of the household work that we are particularly good at voluntary restraint. We do scarcity well when it is enforced and shared, as in the example of water restrictions during drought periods, but are we yet ready to have our propensity for abundance governed – say through rationing? It seems unlikely.

Everyday temporalities

The analysis in this chapter connects to the temporalities of climate change we discussed in Chapter 3. Everyday temporalities can provide sites of both resistance and creativity in responding to the challenges of climate change, particularly for families. In practice these rhythms and movements are subject to disruption and friction, especially where children are involved. Everyday life involves constant attempts to 'smooth' and 'save' time to make routines seamless and minimise interruptions. Certainly the increasing expectations of seamless time and mobility provide points of friction at the moment, and mitigate against a range of more sustainable choices, for example walking instead of driving. Things that enhance flexible timings and save 'wasting' time are relatively non-negotiable; cars are the prime example.

Maintenance of any sense of routine and rhythm in busy schedules is heavily dependent on female labour, usually unpaid. Hence the friction is experienced most strongly by women whose role as household managers involves negotiating, coordinating and integrating the temporalities of family members. In many of the studies in our analysis, a slower pace of household life would be welcomed; it is often the expectations of, and connections with, the wider world that force the speed. But these connections may change by force as the temporalities of modernity unravel.

Some of the transformations we will need to undergo will require us to substitute our own time for fossil fuel dependent activities. Walking rather than driving is the most obvious example, but it is possible to imagine that a whole range of household activities will become more labour intensive. Organo et al.'s (2013) work shows that the transition process and the embedding of new habits take different types of time. These include the *block time* required to research new information or create new infrastructure (build vegetable gardens, organise neighbourhood sharing activities, research installation of solar panels), and the *habitual time* needed to routinise new practices within the household flow (sort rubbish, remember bags so you can shop with less packaging, cook meals from scratch). Households invest time in rethinking their practices at major transition points in the life cycle; when they have a baby, when they renovate, when adult children leave home or when they retire (Gibson et al. 2013). These provide key intervention points for sustainability decisions to be made. Exploring the concept of time helps understand how change occurs – how old practices become deroutinised and new routines embedded. We will probably need to change our experience of time, but this cannot happen in isolation from wider social practices such as work patterns.

In summary, when we think of retrofitting cities and suburbs, attention to the building fabric and infrastructure is a necessary but insufficient condition. We need also to think of retrofitting habits, practices, relations of exchange and sharing. The work discussed in this chapter shows very clearly the diversity of resources that already exists in our communities: fragments of land that can be pieced together, rooftops that can capture energy and water, soil that can be nourished. Those resources include diverse human capacities to make and remake things, share garden produce, space and childcare and to find new patterns of temporality and mobility. Different generations bring different strengths, capacities and knowledge, as do migrants from different parts of the world. These diverse capacities will all be necessary, and provide points of potential leverage as called for by Rowson (2013). They should be documented and celebrated. In a catastrophic world we can expect some reconfiguration of current patterns of vulnerability and strength.

Notes

1 However, detailed accounting of services such as watering, fertiliser use, transport requirements and labour would be required to confirm overall sustainability benefits. It is important to note that we did not attempt to calculate the carbon budget for home-grown vegetables and compare it with commercially grown ones. Home-grown (or local) food is often assumed to be more sustainable because of reduced 'food miles', but research shows that transport is a relatively minor aspect of the carbon costs of food production (Saunders et al. 2006, Roggeveen 2010). It is premature, pending further research, to assume that home-grown food is more sustainable on all criteria. Our intention was simply to assess the potential based on areal calculations.
2 For details of this work, see Waitt et al. (2012), Gibson et al. (2013), Head, Farbotko et al. (2013). An overview of findings and a full list of project publications is provided in Head, Mayhew et al. (2013). The data drawn on in this chapter was compiled as part of a team research project in the Illawarra region of New South Wales. It comprises a representative survey sample of 1465 households undertaken in 2009, a longitudinal ethnography of 16 households undertaken 2010–2011 and in-depth follow-up studies on particular themes with a total of 142 households. In this chapter, where specific findings relate to individual studies, the source is given. For ease of narrative, more general points can be considered to come from the overall findings.
3 While somewhat analogous to the role of meta-analysis in quantitative studies, meta-ethnography is not intended to be deductive, aggregative and averaging. Rather it is faithful to the interpretivist paradigm and the grounded, comparative methods of the original research. The analysis followed the seven broad stages of meta-ethnography identified by Noblit and Hare (1988), as adapted by Britten and Pope (2012). Studies were selected against four criteria: 1) they discussed one or more environmental themes relevant to climate change mitigation and adaptation (without necessarily explicitly discussing them in that context); 2) they are each of a separate household sample; 3) both the qualitative methods and data were reported in the peer-reviewed literature; and 4) both the methods and data were reported in sufficient detail for analysis. The twelve selected studies encompassed 276 separate Australian households, including 158 in the sample from note 2 above. For full details see Head et al. in prep.

References

Abbott, D. and G. Wilson. 2014. "Climate change: Lived experience, policy and public action." *International Journal of Climate Change Strategies and Management* 6(1): 5–18.

Adams, M. 2015. "One blood." *Seminar* 673: 19–23.

Beer, A., S. Tually, M. Kroehn, J. Martin, R. Gerritsen, M. Taylor, M. Graymore and J. Law. 2013. *Australia's Country Towns 2050: What Will a Climate Adapted Settlement Pattern Look Like?* Gold Coast, QLD, Australia: NCCARF.

Blake, J. 1999. "Overcoming the 'value-action gap' in environmental policy: Tensions between national policy and local experience." *Local Environment* 4(3): 257–278.

Blunt, A. and R. Dowling. 2006. *Home*. Abingdon: Routledge.

Brennan-Horley C. and C. Gibson. 2009. "Where is creativity in the city? Integrating qualitative and GIS methods." *Environment and Planning A* 41(11): 2595–2614.

Britten, N. and C. Pope. 2012. "Medicine taking for asthma: A worked example of meta-ethnography." In *Synthesizing Qualitative Research: Choosing the Right Approach*, edited by K. Hannes and C. Lockwood, 41–57. Chichester: Wiley-Blackwell.

Carr, C. and C. Gibson. 2015. "Geographies of making: Rethinking materials and skills for volatile futures." *Progress in Human Geography*, DOI: 10.1177/0309132515578775.

Correa-Velez, I., C. McMichael, S. M. Gifford and A. Conteh. 2015. "Experiences of resettled refugees during the 2011 Queensland floods." In *Applied Studies in Climate Adaptation,* edited by J. P. Palutikof, S. L. Boulter, J. Barnett and D. Rissik, 250–257. Chichester: John Wiley & Sons.

Dietz, T., G. T. Gardner, J. Gilligan, P. C. Stern and M. P. Vandenbergh. 2009. "Household actions can provide a behavioral wedge to rapidly reduce US carbon emissions." *Proceedings of the National Academy of Sciences* 106(44): 18452–18456.

Downes, B. J., F. Miller, J. Barnett, A. Glaister and H. Ellemor. 2013. "How do we know about resilience? An analysis of empirical research on resilience, and implications for interdisciplinary praxis." *Environmental Research Letters* 8(1): DOI: 10.1088/1748-9326/8/1/014041.

Doyle, L. H. 2003. "Synthesis through meta-ethnography: Paradoxes, enhancements, and possibilities." *Qualitative Research* 3(3): 321–344.

Druckman, A. and T. Jackson. 2009. "The carbon footprint of UK households 1990–2004: A socio-economically disaggregated, quasi-multi-regional input–output model." *Ecological Economics* 68(7): 2066–2077.

EPA (United States Environment Protection Authority). 2006. *Protecting Water Resources with Higher-Density Developments*. Washington, DC: EPA.

Farbotko, C. and L. Head. 2013. "Gifts, sustainable consumption and giving up green anxieties at Christmas." *Geoforum* 50: 88–96.

Gandy, M. 2002. *Concrete and Clay: Reworking Nature in New York City*. Cambridge, MA: MIT Press.

Gaston, K. J., P. H. Warren, K. Tompson and R. M. Smith. 2005. "Urban domestic gardens (IV): The extent of the resource and its associated features." *Biodiversity and Conservation* 14(14): 3327–3349.

Ghosh, S. and L. Head. 2009. "Retrofitting the suburban garden: Morphologies and some elements of sustainability potential of two Australian residential suburbs compared." *Australian Geographer* 40(3): 319–346.

Gibson, C., C. Farbotko, N. Gill, L. Head and G. Waitt. 2013. *Household Sustainability: Challenges and Dilemmas in Everyday Life*. Cheltenham: Edward Elgar.

Gibson, C., L. Head and C. Carr. 2015. "From incremental change to radical disjuncture: Rethinking everyday household sustainability practices as survival skills." *Annals of the Association of American Geographers* 105(2): 416–424.

Gibson-Graham, J. K. 2011. "A feminist project of belonging for the Anthropocene." *Gender, Place and Culture* 18(1): 1–21.

Ginn, F. 2012. "Dig for Victory! New histories of wartime gardening in Britain." *Journal of Historical Geography* 38(3): 294–305.

Gleeson, B. 2006. *Australian Heartlands: Making Space for Hope in the Suburbs.* Crows Nest, NSW: Allen & Unwin.

Gregson, N., A. Metcalfe and L. Crewe. 2007. "Moving things along: The conduits and practices of divestment in consumption." *Transactions of the Institute of British Geographers* 32(2): 187–200.

Hasse, D. and H. Nuissl. 2007. "Does urban sprawl drive changes in water balance and policy?: The case of Leipzig (Germany) 1870–2003." *Landscape and Urban Planning* 80: 1–13.

Head, L., C. Farbotko, C. Gibson, N. Gill and G. Waitt. 2013. "Zones of friction, zones of traction: The connected household in climate change and sustainability policy." *Australasian Journal of Environmental Management* 20(4): 351–362.

Head, L., C. Gibson, N. Gill, C. Carr and G. Waitt. In prep. "A meta-ethnography to synthesise household cultural research for climate change policy."

Head, L., K. Mayhew, C. Gibson, G. Waitt, N. Gill, C. Farbotko, N. Klocker and E. Stanes. 2013. *The Connected Household: Understanding the Role of Australian Households in Sustainability and Climate Change.* Wollongong, NSW: Australian Centre for Cultural Environmental Research. Available at http://issuu.com/ausccer/docs/ausccer-connected_household

Head, L. and P. Muir. 2006. "Suburban life and the boundaries of nature: Resilience and rupture in Australian backyard gardens." *Transactions of the Institute of British Geographers* 31(4): 505–524.

Head, L. and P. Muir. 2007. "Changing cultures of water in eastern Australian backyard gardens." *Social and Cultural Geography* 8(6): 889–905.

Heynen, N., M. Kaika and R. Swyngedouw. 2006. *In the Nature of Cities: Urban Political Ecology and the Politics of Urban Metabolism.* Abingdon: Routledge.

Hitchings, R., R. Collins and R. Day. 2015. "Inadvertent environmentalism and the action–value opportunity: Reflections from studies at both ends of the generational spectrum." *Local Environment* 20(3): 369–385.

Hobson, K. 2008. "Reasons to be cheerful: Thinking sustainably in a (climate) changing world." *Geography Compass* 2(1): 199–214.

Houston, D. and K. Ruming. 2014. "Political ecologies of Australian cities." *Geographical Research* 52(4): 351–354.

Howitt, R., O. Havnen and S. Veland. 2012. "Natural and unnatural disasters: Responding with respect for indigenous rights and knowledges." *Geographical Research* 50(1): 47–59.

Inglehart, R. 1997. *Modernization and Postmodernization: Cultural, Economic and Political Change in 43 Societies.* Princeton, NJ: Princeton University Press.

Jackson, T. 2009. *Prosperity Without Growth: Economics for a Finite Planet.* Abingdon: Earthscan.

Kadian, R., R. P. Dahiya and H. Garg. 2007. "Energy-related emissions and mitigation opportunities from the household sector in Delhi." *Energy Policy* 35(12): 6195–6211.

Keogh, D. U., A. Apan, S. Mushtaq, D. King and M. Thomas. 2011. "Resilience, vulnerability and adaptive capacity of an inland rural town prone to flooding: A climate change adaptation case study of Charleville, Queensland, Australia." *Natural Hazards* 59(2): 699–723.

Klocker, N., C. Gibson and E. Borger. 2012. "Living together but apart: Material geographies of everyday sustainability in extended family households." *Environment and Planning A* 44 (9): 2240–2259.

Lane, R. and A. Gorman-Murray. 2011. *Material Geographies of Household Sustainability*. Farnham: Ashgate.

Liverman, D. 2008. "Assessing impacts, adaptation and vulnerability: Reflections on the Working Group II Report of the Intergovernmental Panel on Climate Change." *Global Environmental Change* 18: 4–7.

Loram, A., P. H. Warren and K. J. Gaston. 2008. "Urban domestic gardens (XIV): The characteristics of gardens in five cities." *Environmental Management* 42: 361–376.

Lorenzoni, I., S. Nicholson-Cole and L. Whitmarsh. 2007. "Barriers perceived to engaging with climate change among the UK public and their policy implications." *Global Environmental Change* 17: 445–459.

Mathieu, R., C. Freeman and J. Aryal. 2007. "Mapping private gardens in urban areas using object-oriented techniques and very high-resolution satellite imagery." *Landscape and Urban Planning* 81(3): 179–192.

McCright, A. M. and R. E. Dunlap. 2011. "The politicization of climate change and polarization in the American public's views of global warming, 2001–2010." *The Sociological Quarterly* 52(2): 155–194.

McNamara, K. E. and S. S. Prasad. 2014. "Coping with extreme weather: Communities in Fiji and Vanuatu share their experiences and knowledge." *Climatic Change* 123(2): 121–132.

Mee, K. J., L. Instone, M. Williams, J. Palmer and N. Vaughan. 2014. "Renting over troubled waters: An urban political ecology of rental housing." *Geographical Research* 52(4): 365–376.

Newman, P. and J. Kenworthy. 1999. *Sustainability and Cities: Overcoming Automobile Dependence*. Washington, DC: Island Press.

Noblit, G. W. and R. D. Hare. 1988. *Meta-Ethnography: Synthesizing Qualitative Studies*. Newbury Park: Sage.

Nowak, D. J. and D. E. Crane. 2002. "Carbon storage and sequestration by urban trees in USA." *Environmental Pollution* 116(3): 381–389.

Ockwell, D., L. Whitmarsh and S. O'Neill. 2009. "Reorienting climate change communication for effective mitigation: Forcing people to be green or fostering grass-roots engagement?" *Science Communication* 30(3): 305–327.

Organo, V., L. Head and G. Waitt. 2013. "Who does the work in sustainable households? A time and gender analysis in New South Wales, Australia." *Gender Place and Culture* 20(5): 559–577.

Parker, K., L. Head, L. A. Chisholm and N. Feneley. 2008. "A conceptual model of ecological connectivity in the Shellharbour Local Government Area, New South Wales, Australia." *Landscape and Urban Planning* 86(1): 47–59.

Pauleit, S., R. Ennos and Y. Golding. 2005. "Modeling the environmental impacts of urban land use and land cover change—a study in Merseyside, UK." *Landscape and Urban Planning* 71: 295–310.

Perry, T. and R. Nawaz. 2008. "An investigation into the extent and impacts of hard surfacing of domestic gardens in an area of Leeds, United Kingdom." *Landscape and Urban Planning* 86(1): 1–13.

Peters, G. P., C. L. Weber, D. Guan and K. Hubacek. 2007. "China's growing CO_2 emissions – a race between increasing consumption and efficiency gains." *Environmental Science and Technology* 41: 5939–5944.

Reid, L., P. Sutton and C. Hunter. 2010. "Theorizing the meso level: The household as a crucible of pro-environmental behaviour." *Progress in Human Geography* 34(3): 309–327.

Roggeveen, K. 2010. *Tomato Journeys from Farm to Fruit Shop: Greenhouse Gas Emissions and Cultural Analysis,* Master's thesis. Wollongong, NSW: University of Wollongong.

Rowson, J. 2013. *A New Agenda on Climate Change. Facing Up to Stealth Denial and Winding Down on Fossil Fuels.* Pdf available at www.thersa.org. Accessed October 2015.

Saunders, C., A. Barber and G. Taylor. 2006. *Food Miles – Comparative Energy/EmissionsPerformance of New Zealand's Agriculture Industry.* Lincoln, NZ: Agribusiness and Economics Research Unit, Lincoln University.

Sherrieb, K., F. H. Norris and S. Galea. 2010. "Measuring capacities for community resilience." *Social Indicators Research* 99(2): 227–247.

Shove, E. 2003. *Comfort, Cleanliness and Convenience: A Social Organisation of Normality.* Oxford: Berg.

Shove, E. 2010. "Beyond the ABC: Climate change policy and theories of social change." *Environment and Planning A* 42: 1273–1285.

Smith, R. M., K. J. Gaston, P. H. Warren and K. Thompson. 2005. "Urban domestic gardens (V): Relationships between land cover composition, housing and landscape." *Landscape Ecology* 20(2): 235–253.

Sofoulis, Z. 2005. "Big water, everyday water: A sociotechnical perspective." *Continuum* 19(4): 445–463.

Strengers, Y. 2011. "Negotiating everyday life: The role of energy and water consumption feedback." *Journal of Consumer Culture* 11(3): 319–338.

Strengers, Y. and C. Maller. 2012. "Materialising energy and water resources in everyday practices: Insights for securing supply systems." *Global Environmental Change* 22(3): 754–763.

Timms, P. 2006. *Australia's Quarter Acre: The Story of the Ordinary Suburban Garden.* Carlton, VIC, Australia: Miegunyah Press.

Toole, S., N. Klocker and L. Head in prep. **"**Re-thinking climate change adaptation and capacities at the household scale."

Tudor, T., G. M. Robinson, M. Riley, S. Guilbert and S. W. Barr. 2011. "Challenges facing the sustainable consumption and waste management agendas: Perspectives on UK households." *Local Environment* 16(1): 51–66.

Vale, R. and S. Ghosh. 2006. "Water, water, everywhere ... quantifying possible domestic water demand savings through the use of rainwater collection from residential roofs in Auckland, New Zealand." At the *Urban Drainage Modelling and Water Sensitive Urban Design Conference.* 3–7 April, Melbourne.

Waitt, G., P. Caputi, C. Gibson, C. Farbotko, L. Head, N. Gill and E. Stanes. 2012. "Sustainable household capability: Which households are doing the work of environmental sustainability?" *Australian Geographer* 43(1): 51–74.

Whitehead, M. 2014. *Environmental Transformations: A Geography of the Anthropocene.* Abingdon: Routledge.

Zheng, S., R. Wang, E. L. Glaeser and M. E. Kahn. 2011. "The greenness of China: Household carbon dioxide emissions and urban development." *Journal of Economic Geography* 11(5): 761–792.

10 The Anthropoceneans

What might it mean to inhabit the Anthropocene? What does it mean for us to be citizens of the Anthropocene, both individually and collectively? And how different is that to being a Modern? A key characteristic of the Enlightenment tradition has been that of a hopeful future, the possibility of striving for improvement, both individual and collective. Yet the progressivist view of the future that inspired modernity has helped create the problem. In this concluding chapter I restate the argument and try to braid together the disparate threads of the book by considering the questions above. I have been trying to stretch, and in places invert, our thinking, in order to imagine alternative possibilities. I have done so using everyday materials – practices and examples that are readily to hand in the affluent world. Yet we have to imagine some aspects of these everyday lives being turned on their head. For the purposes of this imagination experiment, we have to assume that sometime in the next few decades, whether by force or choice, we will have to decarbonise so dramatically that many choices we take for granted in contemporary life will no longer be possible.

What is the Anthropocene polity of which we are citizens? If we humans made the Anthropocene, we also made something that is spinning out of our control. So the question of what we are actually citizens of is a live one that we cannot fully answer here. As humans, we are not at the centre of things – the earth does not actually care whether we survive or not. The question of the 'we' is also one that must be specified in different contexts and for different collectives of humans. Nevertheless, let us plunge in, thinking of these Anthropoceneans as the well-off citizens of the Modern world who, having contributed so much to the problems, have to try and remake ourselves and our worlds.

Anthropoceneans have emotional work to do

I have argued that grief and climate change are inextricably entwined. An under-acknowledged process of grieving – with all its complexity, diversity and contradiction – is part of the cultural politics of responding to climate change in the affluent West. Grief and other painful emotions – fear, anxiety, trauma – will be our companion on this journey – they are not something we can deal with once and move on from.

Giving voice to these painful emotions is a somewhat difficult task in a cultural frame with a relentless emphasis on being positive and optimistic as an approach to the future. There is a risk that even discussing such things is somehow not quite 'the done thing'. However, I have come to the conclusion that we are systematically excluding the more extreme parts of future projections from our consideration, just because they are so difficult. If, for example, projected temperature increases for later this century span an envelope between 1.8°C and 4°C, we have no scientific basis for planning for the lower figure just because we like it better and wish it so. That would be really dangerous. My contention is that at least a portion of our preparatory effort must go into emotional preparedness for things that may be extremely unpleasant.

The specific expressions of grief will be variable and much more discussion is needed as to how it can be incorporated more creatively into everyday life. But the evidence from the psychological and psychoanalytical literature is clear: the first step is to acknowledge this companion, grief. If part of what we are grieving for, and what we must farewell, is our modern selves, it follows that a necessary intellectual and practical task is to imagine new kinds of selves.

Anthropoceneans practise hope rather than feel it

If there is work to be done in acknowledging and carrying painful emotions, there is also work to be done in exploring their generative, transformative potential. Anthropoceneans disconnect hope from emotions of optimism, and from an unfolding future. They find hope in practice and being. Disruptive frictions can be welcomed for the opportunities they provide to effect transformation. Prolonged drought has shown the potential to transform water usage practices. Disasters generate networks of care and sharing. In a more catastrophic framing of the future, some of the things that we currently understand as sustainability practices or community support will become survival skills.

Anthropoceneans live in uncertainty without stress

Anthropoceneans need to be able to live in uncertainty. Of course, in many ways human life is always uncertain and fragile, but the loss of a hopeful future means we need to confront our existential uncertainty in different ways. Many of us will need to find home and community in less fixed ways. For example, those of us who live on coasts subject to sea level rise, and shifting boundaries of land and sea, face loss of attachments to place. Or, perhaps, we can build new attachments framed around a more mobile, changing sense of place. We will need to work through important tensions in relation to mobility. On the one hand the hyper-mobilities of late modernity are a key contributor to greenhouse gas emissions; on the other hand sea level rise and other changes mean that whole societies will need to be on the move (Urry 2011).

Further, we have a real evolutionary uncertainty, the knowledge that adapting to today's conditions could bring us undone if those conditions change dramatically tomorrow. Although the broad trends are more than clear, Beven and Alcock (2012) have been reminding us about epistemic uncertainty. Epistemic uncertainty is different to the randomness or the statistical uncertainty in various models. It includes several types of inherent uncertainty; the Rumsfeldian things we don't know we don't know yet, the uncertainties relating to interactions between variables, and the uncertainties arising from non-linear change. We cannot let these uncertainties paralyse us. For Beven and Alcock, 'this means making decisions about the nature of different sources of uncertainty while lacking sufficient information to do so' (2012, p. 5). One scientific response (the wrong one, according to them) has been to try to scale down global models of prediction too far; river X will do this, precipitation will fall this much in region Y in the next few decades, sea level will rise by Z centimetres. Beven and Alcock ask, 'would it not be better to formulate decision-making in a way that does not depend on model predictions?' (2012, p. 4). There are significant challenges here to our governance framings and our understandings of justice, built as they are on notions of stability, predictability and balance.

Anthropoceneans have relational not linear concepts of progress and causation

Anthropoceneans have freed themselves from linear or progressivist understandings of history and progress. If the cost of this is the loss of a hopeful future, there are nevertheless new possibilities in thinking of causality in relational rather than separationist terms. The notion that humans do things to nature, or vice versa, is seen as impossibly oversimplified and empirically inaccurate. Assemblages of many things – some people, some animals, some plants, some ideas, some laws, some bacteria, some atmospheric molecules – are in a constant process of both change and stability. Multiple points of intervention to improve such assemblages (for who? for what?) are possible, indeed necessary, albeit the direction or manageability of change may be out of control. We can have most impact if we attend to amplifying positive change, enhancing ripple effects and working towards tipping points, but it is important to be alert for unacceptable trade-offs and unintended consequences.

We understand that some situations will not be governable, if by governance we are attempting to fix and stabilise conditions. The examples of invasive plants show that we need more mobile and accommodative notions of governance.

Anthropoceneans live in multiple temporalities

Anthropoceneans draw on the heritage of the past, but have no aspirations to return to it. Nostalgia is of little use to us, but knowledge and understanding of the past is extremely useful. If we have set aside a teleological view of progress, we can approach the past in a non-hierarchical way. It is not a state of being we have freed ourselves from, but a resource to help us think about how to do things differently. What kinds of social formations have past communities used to deal with frequent droughts? What kinds of ceremonies bring people together in the face of radical uncertainty? How do you make string from *Typha*? The past provides some imaginative resources to deal with what we currently think of as catastrophe, if we can free ourselves from teleological and progressivist framings of history.

Anthropoceneans respect the long evolutionary history embedded in the various life forms with which we share the Earth. For our own survival as much as theirs, we must maximise the conditions under which diverse life can flourish. This will involve many different kinds of land use and tenure practices. We will need to consider which of our conservationist positions are in fact too conservative, too rooted in preserving the past. Fixed and fenced places that attempt to preserve nature may be stranded as conditions change. Patches, corridors and neighbours will become more important.

This is not a time to remove ourselves from the world. The Anthropoceneans step forward, not back. It is now clear that the metaphor of treading lightly on the earth does not actually help operationalise turning around this Titanic. Uncertainty has led us to frame the debate in terms that conceptually step back: the 're' words reduce/restrain/ restore. These words emphasise concepts of frugality, precaution and localness. These are all good things, and it may be that small and local is where we have to start, but it is not going to be enough. We are going to need things that are transformative, that take courage, that harness abundance.

Anthropoceneans understand how they are embedded in the earth

Anthropoceneans understand the many ways we are embedded – materially, ontologically, historically, biogeochemically – in the processes of the earth. This existential uncertainty is a strangely hopeful source of our understanding. Yes we are different to animals and plants but not in an ontological way – we are connected. And we have certain (different) types of vulnerabilities.

In Australia, research has brought this to our attention more clearly by showing the many traditions of Indigenous people. The combination of palaeoecological and archaeological research, together with the emerging political voices of Indigenous people, challenged the view of pristine wilderness on which many of our conservation ideals rested. In Australia we increasingly understand that we are also embedded spatially, in multifunctional landscapes with many overlapping land tenures and understandings.

The notion of being somehow embedded in the earth makes sense to people. It is widely expressed at the vernacular scale in people's experiences, as documented in these chapters. But caveats are necessary. Sustainable interventions in networks of storage and distribution, such as water, are strongest where people actively understand and participate in those networks. Understanding of 'where things come from' and the networks that support us needs to be fostered. If we are embedded in the earth, environmental management strategies need to acknowledge the pervasiveness of human agency. The networks under discussion need to include the sociality and tenure patterns of human suburban life.

Anthropoceneans understand that their embedding is not all local. Philosopher Val Plumwood (2008, p. 139) put forward the idea of 'shadow places', which bring to light the many unrecognised 'places that provide our material and ecological support, most of which, in a global market, are likely to elude our knowledge and responsibility'. On the positive side, the complexity of global connections means there are also many points of action and political intervention.

Anthropoceneans have ways to articulate and value favourable human action

We have enormous vernacular capacities, not always where we think they are. Documenting and cataloguing such capacities, and the skills and knowledge that underpin them, is an important research activity, as it adds to the menu of future options. Care, stewardship, restoration (but not to a past baseline), humility, planting. We should not focus only on the gentle – vigilance, killing and culling are part of the package. Much environmental work is labour intensive, whether killing invasive weeds in the savanna or juggling household activities to reduce car use. We will increasingly need to recognise and invest in such labour.

Anthropoceneans will have a different orientation to time and its 'use'

As Moderns, we have gradually worked towards seamless time; always on and always effective. This has interacted in particular ways with mobility, such that we can cross the globe in precisely predictable ways. The Holocenean part of us, particularly the agricultural one, works with rhythmic seasonal changes, or their failure. Anthropocene time will have lots of friction, unpredictability, be harder to mediate with technology (or maybe will necessitate new technologies of mediation). We will need to be prepared for interruptions and changes of plan.

We will also need to spend a higher proportion of our days on the necessities of life. As fossil fuel use becomes less possible, and before renewables kick in sufficiently, material standards of living – as currently measured – will surely decrease. Our own labour will need to be substituted for fossil fuel and the cheap labour of distant others. Provision of food and water will probably take more hours of the day, leaving less time for commerce, formal education, cultural pursuits. Or, better perhaps, the provision of necessities will reshape trade, knowledge and culture. We can expect some reduction in the various specialist divisions of labour on which modernity has been built. Mothers of young children may laugh and say that this is what their day looks like anyway – dealing with young people's immediate needs for sustenance.

Shifts in time mean shifts in labour – different kinds of things will need different kinds of labour. Contexts where people are well used to coping with fluctuating and uncertain conditions will come under increasing pressure, and it will not necessarily be clear where the thresholds of coping might be. In Australia, for example, we can ponder the pressures on stretched volunteer capacities as bushfire seasons extend at either end of summer, into spring and autumn.

In times of accelerated change, relationships to tradition can become fraught. We will have to let some things go. Other aspects of the past will become unexpectedly useful; learning to knit from grandma, or learning to fix an engine from grandpa.

The Anthropoceneans are good at sharing

This will not be a utopian equality, as human instincts to provision their own at the expense of the other will surely prevail. And sharing is of course also a quality of a good Modern. There will be a forced reduction in the levels of inequality seen in late modernity, as the capacity to build up huge wealth through global capital disappears. Extreme events will only partially discriminate along axes and gradients of current wealth and capacity. For many, this will mean they have to accept the generosity of strangers in a way they have never had to do before. For others, it will mean sharing things we usually understand in very individualistic terms, like jobs. Jackson and Victor (2011), among others, have argued that a systematic reduction in working hours is one of the ways in which we could make deep emissions cuts without the suffering that would accompany forced recession.

Anthropoceneans live with abundance as well as scarcity

There is considerable evidence now that we do scarcity pretty well, water being the best example. Recent droughts have been an important forcing mechanism. Unfortunately there is less evidence that we do abundance very well, at least in terms of being able to restrain ourselves unless compelled. Perhaps it is helpful to see abundance and scarcity as two sides of the same coin, rather than as states that should be distinguished from one another. We could consider the cross-cutting concepts that help us identify social connections and conflicts. Waste is one point of connection – many people hate waste, even as they have different ideas of what constitutes waste. Sharing is another point of connection – it is expressed in times of flood as celebration and collective storage, and in times of drought as rationing or restrictions.

Australia has been described as having boom and bust ecosystems. They have evolved to cope with huge variability, particularly in the availability of water. This variability is projected to increase under climate change – more intense droughts, more excessive floods. We might want to think more about ourselves and our systems as becoming adapted to such conditions – what might it mean to become a boom and bust society? It's not just about less 'stuff' – sometimes it will be about more 'stuff'.

We are a Holocene subspecies. Our lives have been made possible by agricultural production, and we need to accept this. Agriculture is going to look different in the future, but going back to subsistence farming is not going to be a solution for whole populations. There is a major intellectual task ahead in imagining new kinds of agriculture and horticulture; what bundles of practices can we reconfigure to match more unstable conditions?

Anthropoceneans don't need to be green

The preservationist, conservationist response that has animated many of us for a number of decades is insufficiently powerful to deal with current and projected reality. Our previous conservation ideas and practices do not serve us well enough. On the other hand, many of the capacities we need are ones we already have. They are not necessarily where we think they are, and they do not necessarily have a green or environmental label.

Environmental thought needs to change, but we also have to engage an audience 'beyond environmentalism'. It should not depend on green subjectivities. Rowson argued that the cultural signifiers of the environmental movement were proving a barrier to more widespread engagement with climate change (Rowson 2013, p. 34, see also Roberts 2010). The empirical evidence discussed in Chapter 9 suggests alternative framings that might mobilise more people. Family is one. Family and social relationships are key drivers of household decision-making, even in environmentally conscious households. The family will continue to be central, even as its form may become increasingly diverse.

Anthropoceneans are us

Many of the above qualities are those of good Moderns. The question of how those qualities should be fostered or reshaped will need to be answered differently in different places. My argument has been assembled from a big island on the edge of the world, where the collision of modern and amodern sensibilities has been a source of conflict, but also has great creative and restorative potential. The collision has been responsible for the entrenched social disadvantage of Indigenous people and has brutalised many aspects of the land and its life. Nevertheless it is the hope of many that such ruptures and shards of the past might be part of the raw material from which we craft new futures. Australian evidence provides many challenges to perspectives framed from the northern centres of Enlightenment thought. Human actions in the landscape are evident tens of thousands of years earlier than in other areas. Fire has long been an integral part of ecosystems rather than an external disturbance to them. Deep time is woven into the present. We have a particular cocktail of responsibility for climate change – small population, high per capita emissions, a coal-rich economy. We have particular vulnerabilities, as the most densely populated areas become warmer and drier, extreme events become more frequent and irrigated agriculture becomes less viable. We also have many creative capacities and new sources of strengths in an increasingly diverse population. It is from this particular mix that the antipodean Anthropoceneans can engage, in partnership with other Earth citizens, both human and not. From this and other unlikely places might new models of hopeful Anthropocene life emerge.

References

Beven, K. J. and R. E. Alcock. 2012. "Modelling everything everywhere: A new approach to decision-making for water management under uncertainty." *Freshwater Biology* 57: 124–132.

Jackson, T. and P. Victor. 2011. "Productivity and work in the 'green economy': Some theoretical reflections and empirical tests." *Environmental Innovation and Societal Transitions* 1:101–108.

Plumwood, V. 2008. "Shadow places and the politics of dwelling." *Australian Humanities Review* 44: 139–150.

Roberts, D. 2010. "'Environmentalism' can never address climate change". Accessed 22 August 2015. Available at http://grist.org/article/2010-08-09-environmentalism-can-never-address-climate-change/.

Rowson, J. 2013. *A New Agenda on Climate Change: Facing Up to Stealth Denial and Winding Down on Fossil Fuels.* London: Royal Society for the Encouragement of Arts, Manufactures and Commerce.

Urry, J. 2011. *Climate Change and Society.* Cambridge: Polity Press.

Index

Aboriginal Australians 64–8; colonisation 15; 'domiculture' 98; as 'gardeners' 100; grief 22; landscape burning 133; Protected Areas 47; *Typha* 93; *see also* indigenous people
abundance 3, 95, 151–2, 158, 159–61, 170, 172
Adams, M. 156
adaptation 56–7, 108, 150, 152, 155, 156, 160
agency 58, 60, 61, 64, 124–5; distributed 137, 142; pervasiveness of 170–1
agriculture 13, 15, 44, 93–114, 172–3; biodiversity conservation 121; 'bundles of practices' 98–101, 106–10; emergence of 78; natural rhythms 40; variability 102–6; weeds 116
Ahmed, Sara 14, 22, 24, 77
Albrecht, G. 24
Alcock, R. E. 169
alien species 117–18, 119–20; *see also* invasive species
Anderson, Ben 44, 50n4, 74, 76, 77
Anderson, Kay 15, 43, 58, 62–4, 97–8
anger 77
Angus, Max 38
animals: agency of 124–5; grieving by 24–5; nativeness 15; *see also* non-humans
anthromes 94–5, 100, 120
Anthropocene 8, 15–16, 34, 110; agriculture 94–5, 96, 106; Anthropocene Noir 21; Anthropoceneans 167–73; concept of 4–5; land use policy 133; lines of geographical connection 150; as modernist concept 7, 56; 'new conservation' paradigm 41; posthumanism 68; rapid and unpredictable change 135;

reserve systems 45–7; species extinctions 24; temporality 39–40, 44–5, 49–50, 171; weeds 117
anthropos 7–11, 16
anticipatory grieving 23
anxiety 7, 22, 26–7, 55, 167; Anthropocene Noir 21; climate variability 105–6; Ecological Anxiety Disorder 41–2; pessimism 80; scientists 6, 89
arts 55, 57
Atchison, Jenny 14, 60, 100, 125, 134, 145n1
Australia 13–15, 16n5, 173; agriculture 97–8, 99, 100–1, 103–6, 108–9, 110; archaeological and palaeological evidence 56, 97, 99, 170; autonomy and privacy 157; Blue Mountains 21, 34, 70; boom and bust ecosystems 172; Bundanon 54–5, 70n1, 115, 116, 128; bushfire seasons 172; climate change denial 26, 34n5; colonisation of 43, 62; ethnic diversity 107–8; governance 134, 135–43; humanness 58, 59, 62–8, 69, 97–8; improvement narrative 44; interviews with scientists 80–90; Lake Pedder 38, 39, 50n2; 'Millennium Drought' 158–9; National Reserve System 45–7; native species 120, 121; politics 31; solastalgia 24; suburbs 149; Sydney Basin 62; *Typha* 93–4; Victorian coastal areas 48–9; weeds 115, 120, 121–8; yams 100; *see also* Aboriginal Australians
autonomy 8, 144, 145, 151, 155, 157

Bandt, Adam 21
Banks, Joseph 43
Barker, K. 135

Barton, H. 97
Bastian, M. 47–8
Bawaka Country 58, 64, 66–8, 69
behavioural change 151, 154
belonging 121
Beven, K. J. 169
biocontrol 134
biodiversity 2, 3, 6, 62; agriculture 121; 'Edenic Sciences' 39; Environmental Protection and Biodiversity Conservation Act 136; gamba grass as threat to 138, 144; National Reserve System 46, 47; urban areas 145, 149, 152
Biodiversity in Urban Gardens in Sheffield (BUGS) project 152
biotic nativeness *see* nativeness
Bloch, Ernst 76, 78
Blue Mountains 21, 34, 70
bodies 60, 125
Bonetto, Diego 116
Booth, K. 34n5
boundaries 155
Boyd family 70n1
Brace, C. 44, 45
Bradley, K. 76–7
Brysse, K. 75, 87, 89
Bulkeley, H. 136
Bundanon 54–5, 70n1, 115, 116, 128
'bundles of practices' 95, 98–101, 106–10
bushfires 25, 26, 172; Blue Mountains 21, 34, 70; gamba grass 139, 140–1, 144
Butler, J. 33, 77

Campbell, Barbara 55
Candolle, Alphonse de 118
capitalism 8, 9–10, 12, 55, 96
carbon storage 154
Carr, C. 158
Castree, N. 59
catastrophe 1, 2, 4, 8, 49, 50n4, 133
categories 12
cattle 98
Chakrabarty, D. 58, 61, 67
change 103, 172; attitudes to 156; everyday temporalities 161–2; non-linear 7, 39, 50n4, 106, 133, 169; rapid and unpredictable 135; thresholds for 13
Chase, A. K. 98
Chew, M. K. 118–19, 121
China 15, 150
Chomsky, Noam 34n3

Christmas 160, 161
cities *see* urban areas
Clark, N. 118
Clarke, P. A. 111n2
climate: concept of 6; variability 102, 104–6
climate change 1–4, 11, 13; adaptation 56–7, 108, 150, 152, 156, 160; agriculture 95, 102–4, 110; anxiety and denial 26–7; Australia 14, 173; bushfire risk 21; determinism 44; difference and inequality 60; emotional dimensions of 22, 23, 168; extreme events 25–6; grief 41, 167; households 150; indigenous people 23–4; 'information deficit model' 28; invasive species 133; nativeness 121; reducing consumption 158; relational interventions 61, 145; reserve systems 45–7; society and 56; survey of scientists 80–90; temporality 40, 47–9, 161–2; urban areas 149; variability 172
Cohen, S. 28
colonisation 15, 42–3, 62, 115
compact urban development 152
concepts 12
conflict 127
conservation 62, 109, 170, 173; agriculture 121; Environmental Protection and Biodiversity Conservation Act 136; fortress strategies and fluid strategies 46; 'new conservation' paradigm 41
conservation science 39, 40–1
consumption 15
contingency 7, 45, 97
contradiction 12
Coombes, B. 79
Country 64–5, 69, 70
Crutzen, P. J. 5
cultural diversity 15, 107–8
cultural narratives 29
culture 5, 12, 59, 63, 117
Cunsolo Willox, A. 21, 23, 33

Dark Mountain Project 16n3, 33
Davis, M. A. 119
Davison, A. 117
De Landa, M. 7, 11, 45, 60, 78
De Quincey, Tess 54, 55
defences 27
Denham, T. 97, 98–9
denial 2, 5, 34n6, 42; anxiety and 22, 26–7; collective 1, 6, 14, 23, 32; dealing with denialists 82, 84; as defence against paralysis 85; everyday 48, 85; grief

relationship 11, 32; socially organized 27–9, 88; 'stealth' 29–30
Denisovans 58
destratification 78
determinism 44
Dibley, B. 4
difference 60, 61
disaster recovery 25–6
disavowal 27
Dominey-Howes, D. 26
Doyle, L. H. 155
drought 81, 102, 158–9, 161, 168, 169–70, 172
Duit, A. 133
Dun, Olivia 111n5
Dunlop, M. 46

ecocide 24, 32
Ecological Anxiety Disorder 41–2
ecology 62, 118; *see also* political ecology
economic recession 3
ecosystems 41, 46, 119–20
'Edenic Sciences' 39, 41–2, 50
Ellis, E. C. 94–5, 120
emotional labour/work 75, 83, 88, 167–8; 'compulsory optimism' 86; disaster recovery 25–6; unnoticed 84; use of the term 90n4
emotions 6–7, 11, 22–3, 33, 50, 167–8; denial 28, 30; disaster recovery 26; distancing from 81–5, 88; hope and 74–5; scientists 74, 75, 80–5, 87–8, 89–90
energy consumption 154
energy efficiency 30
Enlightenment 31, 43, 57, 58, 62, 68, 97, 167
environmental management 6, 59, 65, 170–1; *see also* governance
Environmental Protection and Biodiversity Conservation Act (EPBC Act, 1999) 136
environmentalism 6, 38, 173; *see also* green subjectivities
epistemic uncertainty 169
Eriksen, Christine 21
'Erring on the Side of Least Drama' 87, 89
ethnic diversity 14, 107–8
Evans, B. 33
everyday temporalities 48, 161–2
extreme events 25–6, 172, 173

Fagan, B. 96
family 151, 155, 157, 173

Farbotko, C. 22, 77
fear 7, 22–3, 80, 167
Feinberg, M. 32
feminism 22, 60, 89
Fenton, J. 77
Fincher, R. 48–9
fire 9, 15, 70, 173; gamba grass fire risk 139, 140–1, 144; landscape burning 128, 133; *see also* bushfires
flooding 159
Floridi, Luciano 34n3
fluid conservation 46
food, home-grown 162n1
fortress city model 94, 149
fortress conservation 46
fossil fuels 2–3, 31, 32, 58, 171; Australia 14; dependence on 150–1, 161; fortress city model 149; *see also* greenhouse gas emissions
Fritze, J. G. 89
frugality 13, 157, 159, 160, 170
Furberg, M. 23–4

Galaz, V. 133
gamba grass 133, 134, 135–43, 144
Gamble, C. 96
gardens 121–7, 149, 152–4
Gascoigne, J. 31, 43, 44
gender 103; *see also* women
generational continuity 48–9
geoengineering 4, 16n1, 64
Geoghegan, H. 44, 45
geology 7
Ghosh, Sumita 153
Gibbons, A. 58
Gibbs, L. 120
Gibson, C. 10, 14, 68, 77, 158
Gibson-Graham, J. K. 11, 64, 76, 80, 153
Gill, N. 98, 103
Ginn, F. 151
Glendinning, S. 7
Global Financial Crisis (2008) 3
Goodall, Heather 111n5
Gott, B. 110, 111n2
Gould, Stephen Jay 119
governance 13, 128, 133–48, 157, 169
Green activists 87
green subjectivities 13, 157, 173
greenhouse gas emissions 14, 34n5, 150, 168
Grey, G. 100
grief 1–2, 5–7, 21–2, 31–3, 50, 167–8; anticipatory grieving 23; denial

relationship 11, 32; 'Edenic Sciences' 39; hope and 74, 76; non-human loss 24–5; pessimism 80; as a resource 77; speaking about 27; temporality 40–2; *see also* loss; mourning

Hamilton, A. L. 118–19
Hamilton, Clive 31, 34n4, 64, 121
Haraway, Donna 15
Head, Lesley 10, 68, 77, 100, 129n4, 145n1, 153
Hedrén, J. 76–7
Henslow, John 118
herbicide 124, 139
Hinchcliffe, S. 41
history 7, 9, 39–40, 95; Australia 15; non-linear 45, 78, 169; teleological view of 44; *see also* temporality
Hobbs, Richard 40–1, 42, 120–1
Holocene 2, 94, 96, 97, 99, 171, 172
Homo sapiens 9, 58
hope 11–13, 41, 74–92, 168; conceptualising 76–80; as practice 75, 78–9, 88, 90; *see also* optimism
Hornborg, A. 8, 9
households 149, 150–1, 154–6, 157, 161, 173
Howitt, Richie 64–5, 151
Hulme, M. 6, 11, 42, 102, 106
humanism 7, 60, 62, 63–4
humans 10, 12, 57–8, 167; Australian contribution to debates 62–8; human exceptionalism 10, 24, 63; human–nature relations 5, 9, 56, 59, 62–3, 70, 122; humanness and boundaries 117; multiple configurations of the human 58–60; native/alien concept 117–18; relational interventions 61–2
hunter-gatherer societies 78, 95, 97, 101

identity 31, 157, 160
improvement narrative 39, 43–4, 167
India 150
indigenous people 42, 62, 64–8, 69, 79, 110, 170; capacities to cope 156; dispossession 115; emergency management 151; environmental management 59; everyday practices 70; loss of loved places 23–4; National Reserve System 47; rangers 124; social disadvantage of 173; temporality 40; variability 99; weeds 137; *see also* Aboriginal Australians

Industrial Revolution 9, 40
industrialisation 15
inequality 60, 172
'information deficit model' 28
Ingold, T. 97
Instone, Lesley 68, 71n2
Inuit 23
invasive species 117, 119–20, 121; Australian gardeners 121–2, 124, 125, 127; governance 128, 133–45, 169; *see also* weeds

Jackson, T. 34n6, 65, 67, 158, 172
James, S. W. 79

Klocker, Natascha 26, 32, 111n5
Kohn, Eduardo 61
Kondo, T. 101
Koskela, H. 22–3
Kubler Ross, Elisabeth 41

labour 125–7, 171
Lake Pedder 38, 39, 50n2
land degradation 2
land management 140, 141, 142–3, 144
land tenure 107–8, 137, 170
Latour, B. 63
Lavau, S. 43, 46
Leduc, T. B. 10–11
Lien, M. E. 117
Lloyd, Kate 66
Locke, John 43
Lockie, S. 135
Lorimer, J. 10, 41
loss 6, 31; accepting the reality of 33; of loved places 23–4; non-human 24–5; speaking about 27; *see also* grief; mourning
Lulka, D. 10, 60
Lynas, M. 10

Mackellar, Dorothea 16n5, 103, 106
Maller, C. 159–60
Malm, A. 8, 9, 33
Malone, E. L. 3–4
Manderson, L. 25
Marshall, A. 128
Martin, Karen 67
Martin, R. J. 120
Massey, D. 48
materialism 63, 64, 75
materiality 155, 158–61
Mather, C. 128
Mathews, F. 24

Maudlin, T. 59
McGregor, H. 22, 77
McKinnon, C. 23
Mee, K. J. 156
melancholy 74, 76
meta-ethnography 155, 162n3
migrants 107–8, 159, 162
'Millennium Drought' 158–9
Mitchell, Tim 55
modernism 6, 7, 44, 56, 68, 79
modernity 8, 44, 55, 102, 171; grieving for 22, 31; humanist thought 62; ontology of 11; reframing 15; temporalities 39
Mol, Annemarie 78–9
Moore, S. A. 9, 41–2
'more-than-human' thinking 56, 65, 67
mourning 6, 21, 31, 33–4, 38–9, 50; *see also* grief; loss
music 76
mutuality 10, 59

narcissism 26, 27
national reserve systems 45–7
nativeness 15, 41, 117–21, 122, 123
nature 6, 9; framings of human place in 101; human–nature relations 5, 9, 56, 59, 62–3, 70, 122; weeds 115–17
negation 27
Neolithic period 95, 96, 99
'new conservation' paradigm 41
New Zealand 34n5, 120
non-humans 10, 24–5, 60–1, 64, 71n3, 135; *see also* animals; plants
Norgaard, Kari 27–9, 30–1, 48, 85, 88
norms 13, 28, 29, 156
Norway 27–9, 34n5
nostalgia 14, 24, 46, 169
novel ecosystems 41, 119–20

objectivity 83
O'Gorman, Emily 109
optimism 2, 13, 74–5, 88, 168; compared with hope 80–1; 'compulsory' 80, 86, 89; *see also* hope
Organo, V. 161
Oxford Geoengineering Programme (OGP) 16n1

Pahl, S. 40, 44, 45, 47
paradox 12, 155
Parker, K. 145
pastoralism 98, 103, 107–8, 127–8, 133, 138, 143–4
performativity 77, 80

Perrin, C. 63–4
pessimism 22, 34n3, 80–1, 85–6, 87, 88
Phillips, Catherine 60
plants 14, 46, 60–1, 93–4, 115–32; agency of 60, 124–5, 137; 'bundles of practices' 98–101; nativeness 15, 117–21; *see also* invasive species; non-humans; weeds
Pleistocene 8–9, 15, 41, 96, 97
Plumwood, Val 10, 59, 171
policy *see* governance
political ecology 150
pollution 2
posthumanism 10, 57, 59, 65, 67, 68
Preston, C. D. 117–18
priority setting 138, 139–40, 143–4
privacy 151, 157
Prober, S. M. 46
Proctor, J. 9
progressivism 7, 43–4, 97, 167, 169, 170

rain 104–5, 159
Ramankutty, N. 94
Randall, R. 27, 31, 33, 41
Rangan, H. 100–1
rationality 22, 75, 81, 89
Raymond, C. M. 105–6
Rayner, S. 3–4
reasoning 22–3
refugees 107–8, 156
Reid, J. 33
relationality 60, 61, 68, 145, 169
renewable energy 81
reserve systems 45–7
resilience 4, 24, 89, 111n4, 156
resources 13, 156, 157, 162
responsibility 137, 141, 142–3, 144–5
restraint 75
restratification 78
retrofitting 149, 152, 162
Richardson, D. M. 118, 119
Rickards, L. 23
risk assessment 138, 141, 144
Robbins, P. 9, 41–2
Robinson, G. M. 105–6
Roelvink, G. 7
Roeser, S. 22, 23
roof sizes 153–4
Rose, D. B. 21, 42–3, 64
Rowson, J. 28, 29–30, 32, 34n4, 151, 162, 173
Rudd, H. 62

Ruddiman, W. 96
rupture, moments of 77, 78

Sami 23–4
scarcity 3, 95, 151–2, 158, 159–60, 161, 172
sceptics 26, 29, 30, 82, 84, 87
science 5, 7, 28, 50, 55–6, 57, 81; biocontrol 134; debates 41; 'Edenic Sciences' 39, 41–2, 50; emotions in 75, 89
scientists 6, 55, 74, 75; interviews with 80–90; weed management 143
sea level rise 168
seeds 125, 127
self 31, 32, 33
self-reflexivity 89
separationism 56, 109, 122–3, 125
Setterfield, S. A. 133
sharing 172
social context 12
social sciences 5, 55–6, 63, 89, 117, 120
Solander, Daniel 43
solastalgia 24
species declaration 138–40, 143–4
species extinctions 24–5
Spencer, D. C. 75
sprawl 152, 153
'stealth denial' 29–30
Steffen, W. 5, 15
Strengers, Y. 159–60
strengths 156, 162, 173
stress 6, 89
subjectivity 60, 61, 64; *see also* green subjectivities
suburbs 149–50, 152–4, 162
Suchet-Pearson, Sandie 65, 66
Sundberg, J. 57, 65, 67, 69
Sunraysia 104, 107, 110
sustainability 150–1, 154, 157, 161, 168
Sweden 108, 116
Sydney Basin 62

Tasmania 38
Taylor, A. 68
Taylor, M. 56–7, 102–3, 108
Team, V. 25
technology 81
teleology 7, 9, 44, 169, 170
temporality 38–53, 117, 171–2; colonisation 42–3; everyday temporalities 48, 161–2; households 155; multiple temporalities 169–70
'tinkering' 78–9

Tranter, B. 34n5
trauma 25–6, 167
Trigger, D. 120
Truchanas, Olegas 38, 40
Tylor, Edward 43–4
Typha 93–4, 109–10, 128

Ulm, S. 97
'unburnable carbon scenario' 3
uncertainty 5, 7, 11, 133, 168–9, 170; Anderson on 51n4; attitudes to 156; epistemic 169; existential 170; living with 34
United Kingdom (UK) 28, 29, 30, 152, 157
United States of America (USA) 28, 29, 34n5, 70, 77, 157
urban areas 94, 140, 145, 149–50, 152–4
urban ecology 62
urban political ecology 150
Urry, John 31, 55, 78

values 30, 121, 151, 157
Van Dooren, T. 24–5
variability 12, 95, 99, 102–6, 172
vernacular practices 11, 13, 110, 121–8, 145, 171
Victor, P. 172
Victoria 48–9
violence 25, 77
vulnerability 49, 151, 156, 162, 170, 173

Waitt, Gordon 111n5
Walby, K. 75
Warren, C. R. 117–18
water 108, 159–60, 168; gardens 152–3; household demands 154; irrigated agriculture 102, 103, 107, 110; 'Millennium Drought' 158–9; scarcity 161, 172; scientists' views 81; sustainable interventions 170; water degradation 2; water managers 32; *see also* drought
Watson, H. C. 118
weeds 94, 115–32, 171; governance 136–45; nativeness 117–21; practices of living with 121–8; *see also* invasive species
Weeds of National Significance (WoNS) 116, 125, 136, 138, 145
Weintrobe, S. 26–7, 31, 32, 42
Weir, J. K. 24, 32, 67, 68
Western-centrism 8

wetlands 109
Whatmore, S. 65
Whitehead, M. 150
Whittle, R. 25, 84
Willer, R. 32
women 22–3, 97, 99, 161
Wright, Sandra 66

yams 100, 110
Year Zero 42, 43
Yusoff, K. 25, 31, 59, 61, 77

Zalasiewicz, J. 8–9
Zerubavel, E. 28
Zolkos, M. 7